The SAT Math Blueprint

30 Day Plan

Tanjiro Yamamoto

© Copyright Tanjiro Yamamoto 2024 - All rights reserved.

No portion of this book may be reproduced in any form without written permission from the publisher or author, except as permitted by U.S. copyright law.

This publication is designed to provide accurate and authoritative information in regard to the subject matter covered. It is sold with the understanding that neither the author nor the publisher is engaged in rendering professional services. While the publisher and author have used their best efforts in preparing this book, they make no representations or warranties with respect to the accuracy or completeness of the contents of this book and specifically disclaim any implied warranties of merchantability fitness for a particular purpose. No warranty may be created or extended by sales representatives or written sales materials. The advice and strategies contained herein may not be suitable for your situation. You should consult with a professional when appropriate. Neither the publisher nor the author shall be liable for any damages, including but not limited to special, incidental, consequential, personal, or other damages.

Book Cover by Tanjiro Yamamoto

Illustrations by Tanjiro Yamamoto

1st edition 2024

Day-By-Day Overview 6
Day 1 .... *8*
   Introduction and Practice Test 1 8
Day 2 .... *12*
   Arithmetic Basics – Fractions, Percentages, Roots/Exponents 12
   Drill 20
   Drill Explanations 23
Day 3 .... *26*
   Algebra Basics – Simultaneous Equations, Inequalities 26
   Drill 33
   Drill Explanations 35
Day 4 .... *39*
   Quadratic Equations: Grouping, Completing the Square, Quadratic Formula, OSC 39
   Drill 47
   Solving Simple Quadratic Equations Set to 0: Factor the Quadratic into 2 Binomials, and Solve. 49
Day 5 .... *51*
   Word Problems 1 – Rate, Average, Backsolving, Picking Numbers 51
   Drill 61
   Drill Explanations 65
Day 6 .... *71*
   Planar Geometry – Polygons, Circles, 3-Dimensions, Right Triangles 71
   Drill 76
   Drill Explanations 79
Day 7 .... *84*
   Quiz 1 84
Day 8 .... *90*
   Rest, Review, Quiz 1 Explanations 90
   Drill 94
   Drill Explanations 99
Day 9 .... *106*
   Coordinate Geometry – Lines, Parabolas, Rational Functions, Circles 106
   Drill 114
   Drill Explanations 117
Day 10 .... *121*

Word Problems 2 – Representation, Compound Interest 121
Drill... 127
Drill Explanations... 130

Day 11.. 134
Trigonometry and Function Notation – SOH CAH TOA Radians............................... 134
Drill... 138
Drill Explanations... 140

Day 12.. 143
Graphics – Probability, Mode, Median, Probability I, Standard Deviation 143
Drill... 152
Drill Explanations... 157

Day 13.. 163
Quiz 2... 163

Day 14.. 169
Rest, Review, Quiz 2 Explanations.. 169

Day 15.. 177
Break.. 177

Day 16.. 178
Advanced Algebra – Intricate Equations, No Solutions, Infinite Solutions 178
Drill... 184
Drill Explanations... 186

Day 17.. 189
Advanced Geometry – Geo w/Trig, Arcs, Translations 189
Drill... 196
Drill Explanations... 200

Day 18.. 206
Absolute Value, Matching Coefficients, Sequences, Work Rate, Probability II, Statistical Inferences
    206
Drill... 211
Drill Explanations... 213

Day 19.. 216
Quiz 3... 216

Day 20.. 222
Rest, Review, Quiz 3 Explanations.. 222

Day 21..*232*
   Test Experience, Reading and Writing Section, Reviewing Test 1..232
Day 22..*234*
   The Admissions Process...234
Day 23..*238*
   The OSC in Depth, Mental Math, Time-Saving Tips, Nerves..238
Day 24..*246*
   Practice Test 2..246
Day 25..*247*
   Reviewing Practice Test 2...247
Day 26..*248*
   Practice Test 3..248
Day 27..*249*
   Reviewing Practice Test 3...249
Day 28..*250*
   Practice Test 4..250
Day 29..*251*
   Reviewing Practice Test 4...251
Day 30..*252*
   Course Review – Big Picture, Details..252

Day-By-Day Overview

Day 1: Introduction and Practice Test 1

Day 2: We start with a review of basic arithmetic, with an emphasis on fractions, exponents, and roots. That's because much of the exam requires you to manipulate these elements in various ways. If you have any uncertainty about cross-multiplication or negative or fractional exponents, pay close attention to this lesson. You cannot let these basic steps slow you down.

Day 3: Handling simultaneous (multiple) equations is something you're absolutely going to need to do on test day. How are you at substitution, elimination, using the OSC, and choosing among these methods for the most efficient approach?

Day 4: The test people seem to think you can't be a success at college unless you can handle quadratics. These are all over the test. Can you solve through factoring? Can you group? Can you complete the square so you have the vertex form of a quadratic? Do you know the quadratic formula? Can you enter a quadratic into the OSC efficiently? This lesson is a biggie.

Day 5: You need to know the rate and average formulas in their different forms such as "Sum = Average x Number of terms." It's important that you know when to consider backsolving (when the choices are known values) or picking numbers (when the choices contain unknown values such as x).

Day 6: Know your area and volume formulas. Know especially the importance of right triangles, and know the special triangles such as the 45/45 and the 30/60. Other people will know these; you need to know them better.

Days 7 and 8: Quiz 1 and Explanations

Day 9: Mastering this lesson on coordinate geometry will pay off considerably on test day. There's a relationship between algebra and geometry that is spelled out here.

Day 10: The test folks want to know whether you can solve an equation but also whether you understand what the numbers—known and unknown—represent in an equation or expression. And if you don't quite get compound interest, which can show up in various ways on the test, pay close attention to this lesson.

Day 11: There's not a lot of trig on the exam, but knowledge of the basics will be a huge help on some of the tougher questions. Are you still shaky about radians? This lesson will help.

Day 12: You need to be comfortable with the various ways that statistics like mean and standard deviation are represented graphically. And you should be comfortable with calculating the probability of multiple events.

Days 13 and 14: Quiz 2 and Explanations

Day 15: Break

Day 16: This lesson is focused on the trickier algebra that usually appears toward the end of a module. How efficiently can you handle a multi-step algebraic equation? Are you comfortable with the concept of an equation with no solutions, as well as with infinite solutions?

Day 17: This lesson looks at how trig and geometry can be tested in the same question. It also looks at arcs and translations. You might get 1 arc question, and perhaps 1 translation question. Isn't it great to be prepared for it?

Day 18: You'll almost certainly see some "miscellaneous" material on the exam such as sequences and statistical inferences, though probably not all of it. Be ready for anything!

Days 19 and 20: Quiz 3 and Explanations

Day 21: Use this lesson to make sure you have everything you need for test day.

Day 22: Nothing here is directly related to the test since it's about admissions.

Day 23: Learning how to prepare better with quick math and time saving tips will be a huge anxiety reducer on test day

Days 24–29: The test and test review days is where you will solidify your understanding of everything you've learned here.

Day 30: Course review and final words.

Day 1

Introduction and Practice Test 1

*The will to succeed is important, but what's more important is the will to **prepare**.*
—Bobby Knight

After reading this material and completing Practice Test 1, sign and check the box for Day 1.

The SAT is a wonderful thing.

It's wonderful because you can succeed at it without **luck**. This is not true with everything in life, is it? Where you were born, when you were born, who you happen to meet along the way—they're all out of your control. And being lucky sure is nice.

Getting a great score on the SAT, however, is **in** your control. For example, the test does not:

- ☐ determine your "natural aptitude."
- ☐ measure your ability to do complicated calculations. (If you find yourself multiplying 3,289 by 413, you're almost certainly doing something wrong.)
- ☐ give a particularly accurate account of how well you've done in math class. Here is one of the most common complaints we hear from students (and their parents): *"I do great in school, so why isn't that reflected on my SAT score?"*

Here's what the SAT **does**:

- ☐ It tests whether you have a working knowledge of a fairly small set of principles that are found again and again on every test, year after year. If you've made it to your junior year of high school (or beyond), you have what it takes. Now you just need a plan.

Now you have that plan. If you fully commit, if you make this experience a priority, you will be ready for pretty much anything the test throws at you. From the very first question, you'll be thinking, "I've seen this before. I know what to do here. I can handle this." In a word, you'll be **prepared.** And that's a wonderful feeling. There's a reason that **being unprepared for a test** is one of the most common nightmares. It's a scary feeling. Once you complete this 30-day action plan, you might not love sitting for the exam, but you should not be scared of it.

Why This Book?

Deadlines work. That's one reason for this book.

The science is clear. If you want to accomplish a task, then having a plan with **specific deadlines** will increase the likelihood of success many times over. Knowing that we need to do **this** today and **something else** tomorrow in order to get to a highly desirable outcome on **this particular date** triggers powerful psychological forces in us. It programs us for success. Top athletes know this, virtuoso musicians know this, and business leaders know this. You might say, "Well, doesn't everybody know this?" Maybe, but very few actually put this powerful principle into action. It takes big-picture thinking and long-term, highly organized planning.

Planning is what we've done for you. We've got your plan.

And **that's** what makes this book different from all the other test prep books on the market. Sure, we know this test inside-out, and we know the areas where students struggle. And we're not the only ones who do. **But the big difference between us and the others** is that our focus is on **you**. We've worked with thousands of students, and we know that even highly motivated test-takers don't always know what to do **now** and what to do **next**. From Day 1 to Day 2 and right up to Day 30 in this book, you will have specific tasks that build on each other. As you check off the boxes we provide, you'll be filling in your "math gaps" so you'll be an SAT expert on the day of the test.

We provide the tools, we point the way, we guide you step-by-step toward your goal, but it's **up to you to do the work**. There's a difference between a student whose score flatlines after prepping for the exam and a student who gets the score they truly desire (or better). The student with the high score had a plan and **then executed it**.

Isn't checking off a list of things you need to do a great feeling? Make sure you do exactly that with this course. You might also tell others what you're doing to help motivate you to do your best. **Perhaps you know another test-taker you can share your day-to-day accomplishments with.**

We've spent a great deal of time designing this course so 30 days is **exactly** what you need. Unlike other courses, this one is focused like a laser beam on the skills you'll need for the SAT—**nothing more, nothing less**. Each day builds on the previous one. One step at a time, **you will get there!**

Your first concrete step is to take a full-length, carefully timed Math section test. (See the **note** at the end of this Introduction.) The practice test serves 2 purposes:

1. You'll get familiar with taking a digital SAT.
2. You'll have a better idea of where to concentrate your energies.

Jumping to the end of the 30 days, the final steps also involve taking full-length Math section tests. You'll see how far you've come, and you'll see those areas that could still use your attention. Every question you work on is a golden opportunity, a chance to *not* get a question of that sort wrong on the actual test. You probably won't be happy about fumbling a question, but think of it this way: "Great! I've found a problem I can fix before test day."

Introduction and Practice Test 1 | 9

Between the first and last steps of the 30-day plan, you'll be mastering—one by one—all the math content found on the test. You'll also learn to approach the material strategically. And as you master the content and learn the strategies, you'll automatically be taking care of that all-important factor—**time management**. Have you ever wished you had more time on a math exam? We're here to give it to you. If you spend the next month bearing down on this material, you will **have more time** on the test because you'll handle the questions—including the most challenging ones—with maximum **efficiency**. That's the key word. It all comes down to the most **efficient** way to handle a question.

Thirty days—is that enough? You might know some people who spend **much more time** than that prepping for the SAT, but that doesn't mean they're getting **more prepared**. And let's face it. There are other things you could be doing with that time, including getting your GPA up and becoming involved in activities that college admission people love to see on applications. So the short answer is yes. Thirty days is enough time if—

—**if** you follow the plan, do the work, and stay focused. Like it or not, the SAT is a big deal. It will help decide where you're going to college, who you'll meet, and where you will go from there. This is one of the rare times in your life when putting in a month of hard work will have **lasting, positive effects**. There's no luck required. You want to succeed, and **we** want you to succeed. Together, we can do this.

Let's get started.

Today, you'll take the first of four official Digital SAT Practice Tests directly on the College Board site using your laptop or tablet.[1] Then take a look at the results. You'll be looking at this test again toward the end of this course. It is important that *all practice tests are taken through the College Board site*, as these are the only official practice tests and will most closely mirror what you will see on test day.

You'll also want to **use your emotional reaction to your score** to launch you into this course. Did you run out of time? Did you forget crucial formulas? Did you have **no idea** how to handle certain questions? If the answer is yes to at least 1 of these questions, then let that motivate you to get everything you can out of this course. Don't skip a day. Make sure you take all the drills and all the quizzes. Know the formulas!

The Digital SAT

While the math **content** on the digital version of the test is similar to the content on the paper exam, the **format** has changed. Here's what you need to know.

- ☐ You'll get two 35-minute Math sections—called *modules*—each with 22 questions. (These come after the two Reading and Writing modules and a short break.)

---

[1] You can find a paper copy of these tests, but, unless you're someone who needs to take it that way on the day of the actual exam, you should stick to the online practice tests.

- [ ] The test is *adaptive*. The first Math *module* will contain questions of medium difficulty. If you do well on this module, the second module will be more difficult. If you don't do so well, the second module will be easier.

- [ ] Correctly answering questions on the harder module will increase your score **more** than correctly answering questions on the easier module. Therefore, you **want** the harder module.

- [ ] Like the paper test, the digital math questions gradually get more difficult as you progress through each module. And by the end of this course, you'll know shortcuts that can actually make the more difficult questions quite simple.

- [ ] Like the paper test, most of the digital questions have 4 choices, although some are student-produced response (SPR) questions that require you to type in the answer.

- [ ] You can bring your own calculator, but a graphing calculator is available on the screen for every question. If you aren't familiar with this calculator, **you need to be**! This device is significantly easier to use than an ordinary calculator and can often be very useful during the test. We'll refer to it as the on-screen calculator (**OSC**) and will look at in depth during the course.

- [ ] You can go back and change an answer at any point during each timed module.

- [ ] You'll have access to some geometry formulas on the screen.

- [ ] You can mark a question to remind yourself to check it again if you have time.

- [ ] You'll be given a few sheets of paper to do your work on.

- [ ] You'll have shortcut commands for a few simple actions such as zooming in or out.

Note: Should you take the Reading and Writing section today? It depends.

You'll be taking 4 official math tests during this course, including today's. We recommend that on at least 1 of those days, you also take both modules of the Reading and Writing section. If you are highly confident in your Reading and Writing skills, then you may only want to take that module once. Otherwise, you may want to consider taking the full test with both modules all four times. It will be important to test your stamina and timing by sitting through the full-length exam exactly like you will on test day.

If you are only taking the Math sections, you'll need to bypass the two Reading and Writing modules to get there. The easiest way to do that is to skip every question on those modules so it will take you to the Math section. Yes, it will look like you didn't get any questions right on the Reading and Writing section, but no one will ever know.

Day 2

Arithmetic Basics – Fractions, Percentages, Roots/Exponents

After reading this material and completing the drill at the end, sign and check the box for Day 2.

It's time to make sure you have command over exactly those fundamental skills you **must** master in order to ace this test. After thoroughly reading the following material, you'll complete a drill to find out if you really know this stuff. Having these skills at the ready is no guarantee you'll attend the school of your dreams, but without them—well, you're doomed. In other words, it's time for the basics.

The Basics (and Therefore the Most Important Stuff)

Let's quickly review some vocabulary. You probably know what integers are. Some people call them *whole numbers*, but it's important to know that even -2 and 0, for example, **are integers** just like 7, 10, 50, and more. That's simple enough, but students sometimes make mistakes because they don't absorb the part of a question that says, for example, "x and y are integers." A question will be impossible to solve if you automatically think *positive integers* and forget about 0 and the negatives.

Fractions

Fractions are the numbers between the integers. Fractions are a big deal on the SAT. You will see them in many questions, and **that's a wonderful thing** because you'll soon be an absolute master of working with fractions. Many students would rather deal with decimals since calculators are more decimal-friendly. But most of the time, you're better off working with fractions, and that's what you'll mostly see on the exam.

You Must Remember This
No matter how complicated fractions look, stick to the basics below, and you'll be fine.

Adding/Subtracting Fractions: Find a common denominator (preferably the least common denominator).

Simple: $\frac{1}{2} + \frac{2}{3} = \frac{3}{6} + \frac{4}{6} = \frac{7}{6} = 1\frac{1}{6}$

In this example, the least common denominator is 6.

Challenging: $\frac{3b+c}{3d} - \frac{5c}{4e}$

The common denominator is the product of the 2 denominators: 12*de*. Therefore, you'll have to multiply the top and bottom of the first fraction by 4*e* and then multiply the top and bottom of the second fraction by 3*d*, as shown below.

$$\frac{3b+c}{3d} - \frac{5c}{4e} = \frac{4e(3b+c)}{12de} - \frac{3d(5c)}{12de} = \frac{12be + 4ce}{12de} - \frac{15cd}{12de} =$$

$$\frac{12be + 4ce - 15cd}{12de}$$

Here's what you should tell yourself: "I don't care what it **looks like**. If I'm adding or subtracting fractions, I'll find a common denominator."

What goes wrong/what takes too long: Confusing it with **multiplying** fractions.

Multiplying Fractions: Multiply across the top and then the bottom. Cancel or simplify when you can.

Simple: $\frac{3}{5} \times \frac{1}{2} = \frac{3}{10}$

Challenging: $\frac{5x}{y} \times \frac{3w+1}{15z} = \frac{5x(3w+1)}{15yz} = \frac{15wx + 5x}{15yz}$

This can be simplified by dividing the top and bottom by 5.

$$\frac{15wx + 5x}{15yz} = \frac{3wx + x}{3yz}$$

(If you canceled or simplified at an earlier stage, that's fine.)

What goes wrong/what takes too long: Confusing it with addition/subtraction. It also might be confusing multiplication with cross-multiplication (we see this a lot). We'll get to that shortly, but remember, we only **cross-multiply** when we have an **equation**. If you don't have an equal sign, you cannot cross-multiply.

Dividing Fractions: Invert the bottom, and then multiply.

Simple: $\dfrac{\frac{1}{2}}{\frac{3}{5}} = \frac{1}{2} \times \frac{5}{3} = \frac{5}{6}$

Challenging: $\dfrac{\frac{6r+s}{5t}}{\frac{t}{2}} = \frac{6r+s}{5t} \times \frac{2}{t} = \frac{12r+2s}{5t^2}$

Arithmetic Basics – Fractions, Percentages, Roots/Exponents | 13

What goes wrong/what takes too long: Forgetting to invert the bottom fraction. Also, students sometimes get confused when dividing a fraction by an integer. For example, in $\frac{\frac{7xy}{z}}{5} = \frac{7xy}{z} \times \frac{1}{5} = \frac{7xy}{5z}$, note that 5, when inverted, becomes 1/5.

Cross-Multiplying with Fractions: As mentioned above, you can only cross-multiply with fractions when you have an equation. Multiply the top left with the bottom right, and **set that equal to** the product of the top right and the bottom left. Then you can isolate and solve for the unknown value.

$$\text{Simple: } \frac{5}{7} = \frac{x}{10} \rightarrow 50 = 7x.$$

Now we can divide both sides by 7: $\frac{50}{7} = x$. This can also be written as $x = 7\frac{1}{7}$.

$$\text{Challenging: } \frac{7a+2}{5} = \frac{2a}{9} \rightarrow 63a + 18 = 10a.$$

By collecting like terms and isolating a, we get $a = -\frac{18}{53}$.

What goes wrong/what takes too long: The following happens **a lot** but won't happen to you! Students will multiply (correctly) the top left with the bottom right and then (incorrectly) place that **over** the product of the top right and the bottom left, ending up with a fraction. That makes no sense because you no longer have an equation. You must have an equal sign after cross-multiplication. You can then isolate the unknown (x or whatever) on one side of the equation.

Splitting a Fraction: This is the opposite of adding/subtracting. In a sense, you're complicating rather than simplifying, but in some problems, that will be helpful.

$$\text{Simple: } \frac{3a+b}{x} = \frac{3a}{x} + \frac{b}{x}$$

$$\text{Challenging: } \frac{3a-b}{4a} = \frac{3a}{4a} - \frac{b}{4a} = \frac{3}{4} - \frac{b}{4a}$$

What goes wrong/what takes too long: Splitting can be done when the top (*okay, we'll call it by its proper name—the numerator*) involves addition/subtraction but **not** when it only involves multiplication. For example, $\frac{3abd}{5c}$ cannot be split.

Using Fractions to Express a Ratio: The word *to* tells you where the fraction bar goes, as does the colon sign (:).

$$\text{Simple: "The ratio of } x \text{ to } y \text{ is 3 to 5" becomes } \frac{x}{y} = \frac{3}{5}$$

Challenging: This isn't particularly challenging,
but you might see "the ratio of the shortest side of a triangle to the longest side is 2:7."
If you then use l to represent the longest side and s to represent the shortest side,
you can express this as $\frac{s}{l} = \frac{2}{7}$.

What goes wrong/what takes too long: Some students just don't know (or remember) to use a fraction to express a ratio. It's almost always a good idea because then you can cross-multiply or do whatever you need to do next to solve the problem.

Converting to a Decimal or Percentage: Memorize some of the common ones, and use your calculator for others.

Simple: You should know the decimal equivalents of 1/10 through 9/10,
1/5 through 4/5, thirds, and of course ½.

Challenging: eighths, sixths
If you use your calculator to, for example, determine that 7/8 is .875,
move the decimal point over twice (this is the same as multiplying by 100),
and then stick on a percentage sign (which is the same as dividing by 100).
So, 7/8 = .875 = 87.5%. And that leads us to:

Wait! Wait! Before moving on, here's one thing.

Percentages

Finding a Percentage of: This is the same as multiplication.

Simple: 30% of 15
You can easily convert 30% into .3 and then multiply it by 15, giving you 4.5.
You could also write this as 3/10 x 15—same thing, same answer,
although it might **look** different.
Using 3/10 will give you an answer of 9/2, which is the same as 4½ or 4.5.
Any of those could be given as the correct choice.

Challenging: 122% of 10% of 40
Since you're multiplying these numbers, it doesn't matter which you do first.
You might first determine 10% of 40—which is 4—and then
multiply that by 122% (1.22).
That gives us 4 x 1.22 = 4.88.

What goes wrong/what takes too long: Writing, for example 45% as 4.5 instead of .45. If you're converting to a decimal, watch where you put the point. Also, don't confuse "finding a percentage of" with the following.

This Is What Percentage of That? This also uses the word *of* but not in the same way as the examples above. This is actually division and can be expressed as a fraction.

Simple: 15 is what percentage of 120?
You can express this as 15/120.
Simplifying this (with or without a calculator) gives you 1/8 or .125, which is 12.5%.

Challenging: 27 is what percentage of 2?
(Right away you know the answer must be more than 100%.)
This is 27/2, which is 13.5 or 1350%.

What goes wrong/what takes too long: Confusing this with the following.

Percentage Greater Than/Less Than: This important concept can be expressed several ways and usually involves more than 1 step. You'll find the actual difference between the two values and then put that difference over the number that comes after **that.** Then you can convert this to a percentage.

Simple: 11 is what percentage **greater than** 8?
Step 1: Find the actual difference, which is obviously 3.
Step 2: Place this difference over 8, giving you 3/8 or .375 or 37.5%.

Here's another example: 20 is what percentage **less than** 30?
Step 1: Find the difference, which is 10.
Step 2: Place this difference over 30,
giving you 1/3 or .333 (repeating) or 33 1/3%.

Challenging: The cost of a jacket increased by 10% and then again by 20%.
The final price is what percentage more than the original price?

Be careful with this one. The answer is not 30%, which would be too simple for the test. When you are not given a starting value, it's often a good idea with percentage questions to **choose 100**. So we'll proceed as if the original cost of the jacket was $100.

Step 1: Find the value that is 10% **more than** 100, giving you 110.
Step 2: Find the value that is 20% **more than** 110.
Since 20% **of** 110 is 22, we add that to 110 to get 132.
This is the final price: $132. But we're not done.

The question was this: "The final price is what percentage more than the original price?" Therefore, we find the actual difference (132 − 100 = 32) and put this over the original price, so we get 32/100, which is 32%.

What goes wrong/what takes too long: Students often take too many steps for this kind of problem. In fact, we did exactly that in the explanation above. Our Step 2 above was to "find the value that is 20% **more than** 110." This can be done in 1 step. Rather than finding 20% of 110 and then adding it on, just multiply 110 by 1.2. In another example, if you want the number that is 15% more than 70, just multiply 70 *not* by .15 but by 1.15, giving you 80.5.

Another **shortcut** to keep in mind is that, for example, 13% of 100 is 13. 243.6% of 100 is 243.6. That's one reason for choosing 100 as a starting value if they don't give you one.

Roots and Exponents

A lot of roots and exponents will show up on the exam. If you have a phobia about them, it's time to fix that.

Exponents: As you no doubt know, $7^2 = 49$. What about 7^3? Well, you have a calculator for that, but it's unlikely that you'll face this on the test because the good people who create the SAT **know you have a calculator.** They're much more interested in seeing if you understand the way exponents can be used, manipulated, and so forth. Let's look at some examples.

Multiplying with Exponents: This can only be done when **the bases are the same.** You keep that base and add the exponents.

$$\text{Simple: } 7^2 \times 7^3 = 7^5$$

$$\text{Challenging: } 7^{x+1} \times 7^{2x} = 7^{3x+1}$$
Is that really that much more challenging?
Not if you keep in mind the golden rule: If the bases are the same,
keep that base, and add the exponents.

Dividing with Exponents: This is just the opposite. If the bases are the same, keep the base, and subtract the exponents. That will mostly come up with fractions.

$$\text{Simple: } \frac{7^5}{7^3} = 7^2$$

$$\text{Challenging: } \frac{x^{(3+2y)}}{x^y} = x^{3+y}. \text{ Now is } \textbf{that} \text{ challenging?}$$
Well, it's potentially **confusing.** But all you're doing is keeping that base (x)
and then subtracting the exponents $(3 + 2y) - y = 3 + y$.

What goes wrong/what takes too long: One obvious mistake is to **multiply** the exponents when you should add them (or divide them when you should subtract). Here's a typical case where that could happen.

$$x^r x^s x^t = x^{r+s+t}$$

This equation is correct. However, many students will mistakenly multiply the exponents, including those students who **know the rule.** People just don't like to see complicated-looking exponents like this one. Don't be those students!

Raising a Power to a Power: Multiply the powers.

$$\text{Simple: } (x^5)^3 = x^{15}$$

$$\text{Challenging: } [(a)^{b+c}]^d = a^{bd+cd}$$
Note that the entire exponent $b + c$ gets multiplied by d.
You need to *distribute* the d.

Arithmetic Basics – Fractions, Percentages, Roots/Exponents

Negative Exponents: Invert the base, and then make the exponent positive. That's it.

$$\text{Simple: } 7^{-2} = \left(\frac{1}{7}\right)^2 = \frac{1}{49}$$

Note that the result of raising the base (7) to a negative exponent does **not** make the result negative.

$$\text{Here's another: } \left(\frac{1}{3}\right)^{-3} = 3^3 = 27$$

Again, there are 2 steps.
Invert (turn upside-down) the base, and make the exponent positive.

$$\text{Challenging: } \left(-\frac{2}{5}\right)^{-2} = \left(-\frac{5}{2}\right)^2 = \left(-\frac{5}{2}\right) \times \left(-\frac{5}{2}\right) = \frac{25}{4} = 6\frac{1}{4} = 6.25$$

Note: We ended up with a positive value since we multiplied 2 negatives.

What goes wrong/what takes too long: Thinking you need to end up with a negative value just because you've got a negative exponent. Nope. As for **time-savers**, have a look at the following equation with negative exponents in a fraction.

$$\frac{x^{-3}y}{x^2 y^{-1}} = \frac{y^2}{x^5}$$

Note what happened. The *x* raised to the negative third got dropped down to the bottom. Its exponent became **positive** 3, and then when this was multiplied by *x* squared down there at the bottom, we got *x* to the 5th. Likewise, the *y* raised to the negative 1 on the bottom got moved up to the top, and its sign was changed to positive 1. When this was multiplied by the *y* already up there, we get *y* squared. This is all based on our rule of inverting and sign-changing, but it's a useful shortcut.

Fractional Exponents: These are the same thing as roots.

$$\text{Simple: } 25^{\frac{1}{2}} = \sqrt{25} = 5 \text{ OR } 27^{\frac{1}{3}} = \sqrt[3]{27} = 3$$

(More on roots are coming up soon.)

$$\text{Challenging: } 27^{\frac{2}{3}} = \left(\sqrt[3]{27}\right)^2 = 3^2 = 9$$

In other words, we find the cubed (3rd) root of 27, which is 3 (because 3 cubed is 27). Then raise that to the 2nd power, giving us 9. Let's look at another one.

$$36^{\frac{3}{2}} = \left(\sqrt{36}\right)^3 = 6^3 = 216$$

Since we raised 36 to a power **greater than 1**,
we end up with a number greater than 36.

What goes wrong/what takes too long: Too many students just forget about the basic rule and when confronted with $100^{\frac{1}{2}}$ will end up with 50. But not you.

One more thing—well, 2 more—before we move on. Let's talk about nothing—that is, 0. When 0 is raised to a power, like 0^{17}, we get 0. However, anything raised to the 0 power is 1. So $17^0 = 1$. But 0^0 is undefined. There's no such thing.

Roots: The opposite of exponents.

$$\text{Simple: } \sqrt{121} = 11$$

Challenging: $\sqrt[5]{32} = 2$. This is not really all that challenging.
What would be challenging is if you were asked for, say,
the 11th root of 562, but that's not what the test is about.
They want to know if you understand the **concept**.

What goes wrong/what takes too long: You might have the right answer, but it's not in the right **form.** You might need to do the following.

Factoring Roots: Some roots can be expressed as the product of 2 numbers, and 1 of them is an integer.

$$\text{Simple: } \sqrt{50} = \sqrt{25} \times \sqrt{2} = 5\sqrt{2}$$

Note that we factored root 50 into 2 numbers, and 1 was a **perfect square.**
And 25 is a perfect square, meaning that its square root is an integer.
We could have instead factored it this way: $\sqrt{50} = \sqrt{10} \times \sqrt{5}$.
But there's nowhere to go from there.

Challenging: These won't get much more challenging than the example above.
But here's something to keep in mind.
One way to get comfortable with roots is to approximate their value.
Take $\sqrt{50}$. It's obviously a little more than $\sqrt{49}$.
Since that's 7, we know that $\sqrt{50}$ is just a very little bit more than 7.

What goes wrong/what takes too long: Root of numbers that aren't perfect squares are **irrational numbers.** So the square roots of 3, 5, 7, 8, 10, and so forth are all irrational. You cannot write them accurately as a fraction or a decimal. In decimal form, they go on forever **without a pattern.** Sometimes students will get, say, $\sqrt{103}$ as their answer and think that's not good enough. No, that's as good as it gets. Take a look at the answer choices. There it is!

One thing to keep in mind about irrational numbers is that we typically are not supposed to divide by them. So if you get $\frac{8}{\sqrt{2}}$ as an answer, you might be right, but it's not in the right form. So we multiply that fraction by $\frac{\sqrt{2}}{\sqrt{2}}$. We get $\frac{8}{\sqrt{2}} \times \frac{\sqrt{2}}{\sqrt{2}}$, which equals $\frac{8\sqrt{2}}{2}$, which simplifies to $4\sqrt{2}$. It's the same answer but in a different form.

Bonus Note: We've said that exponents and roots are the opposite of each other. While that's true, we have to be careful about the following: $x^2 = 16$ has 2 solutions—4 and negative 4. However, the square root of 16 is only 4. The square root sign means "the positive root."

And now it's time for a drill that reviews all the previous material.

Arithmetic Basics – Fractions, Percentages, Roots/Exponents | 19

Day 2

Drill

*If you are confident about a topic below, you can do only the **bolded** questions. In any case, aim for **efficiency**! Be **organized**! Write your work neatly, and only write what you need. We know this will help!*

Fractions

Adding/Subtracting Fractions: Find a common denominator.

1. $\frac{2}{9} + \frac{1}{6} =$
2. $\frac{7r}{st} - \frac{2}{t^2} =$

Multiplying Fractions: Multiply across the top and then the bottom. Cancel/simplify when you can.

3. $\frac{3}{7} \times \frac{5}{a} =$
4. $\frac{ab^2}{c} \times \frac{a+1}{c} =$

Dividing Fractions: Invert the bottom, and then multiply.

5. $\frac{\frac{4}{5}}{\frac{2}{3}} =$
6. $\frac{\frac{2}{7a}}{b-1} =$

Cross-Multiplying with Fractions: Multiply the top left with the bottom right, and **set that equal to** the product of the top right and the bottom left. Then you can isolate and solve for the unknown value.

7. $\frac{y}{9} = \frac{2}{3}, y =$
8. $\frac{x}{2} = \frac{3x}{y+1}, x \neq 0, y =$

Splitting a Fraction: This is the opposite of adding/subtracting.

Split the following fractions into 2 parts. Simplify if you can.

9. $\frac{2+x}{3} =$ 10. $\frac{3a-ab}{b} =$

Using Fractions to Express a Ratio: The word *to* tells you where the fraction bar goes, as does the colon (:).

Translate the following into equations with fractions.

11. The ratio of *x* to *z* is 17:20. _____

12. The ratio of the length (*l*) of a rectangle to its width (*w*) is 5 to 6. _____

Converting to a Decimal or Percentage: Memorize some of the common ones, and use your calculator for others.

Express the following fractions as decimals *and* as percentages.

13. $\frac{3}{5} =$ 14. $\frac{5}{8} =$ 15. $\frac{4}{3} =$

Percentages

Finding a Percentage *of*: This is the same as multiplication.

16. 200% of 16 = **17. 1/2 % of 50 =**

***This* Is What Percentage of *That*:** This is actually division and can be expressed as a fraction (or a decimal), which can then be expressed as a percentage.

18. 12 is what percentage of 18? **19. 2 is what percentage of 1/2?**

Percentage Greater/Less Than: This important concept can be expressed several ways and usually involves more than one step.

20. 12 is what percentage greater than 10? 21. 15 is what percentage less than 100?

22. The price of a shirt increased by 5% and then increased by 10%. The final price is what percent more than the original price?

23. The area of a circle is increased by 20% and then decreased by 20%. The final area is what percent of the original area?

Drill | 21

Roots and Exponents

Multiplying with Exponents: This can only be done when **the bases are the same.** You keep that base and add the exponents.

24. $y^3 \times y^8 =$

25. $r^2 r^3 r^4 =$

Dividing with Exponents: If the bases are the same, keep the base, and subtract the exponents.

26. $\dfrac{x^5}{x^4} =$

27. $\dfrac{x^5}{x^{-2}} =$

Raising a Power to a Power: Multiply the powers.

28. $(x^5)^3 =$

29. $(y^{-3})^{-10} =$

Negative Exponents: Invert the base, and then make the exponent positive.

30. $\left(\dfrac{1}{2}\right)^{-3} =$

31. $\dfrac{ab^{-3}c}{bc^{-3}} =$

Fractional Exponents: These are the same as roots.

32. $81^{\frac{1}{2}} =$

33. $64^{\frac{2}{3}} =$

Roots: The opposite of exponents.

34. $\sqrt[3]{8} =$ 56.

35. $\sqrt[4]{81} =$

Factoring Roots: Some roots can be expressed as the product of 2 numbers, 1 of which is an integer.

Factor the following.

36. $\sqrt{40} =$

37. $\sqrt{200} =$

Day 2

Drill Explanations

Fractions

Adding/Subtracting Fractions:

$$1.\ \frac{2}{9}+\frac{1}{6}=\frac{4}{18}+\frac{3}{18}=\frac{7}{18} \qquad 2.\ \frac{7r}{st}-\frac{2}{t^2}=\frac{7rt}{st^2}-\frac{2s}{st^2}=\frac{7rt-2s}{st^2}$$

Multiplying Fractions:

$$3.\ \frac{3}{7}\times\frac{5}{a}=\frac{15}{7a} \qquad 4.\ \frac{ab^2}{c}\times\frac{a+1}{c}=\frac{a^2b^2+ab^2}{c^2}$$

Dividing Fractions:

$$5.\ \frac{\frac{4}{5}}{\frac{2}{3}}=\frac{4}{5}\times\frac{3}{2}=\frac{12}{10}=1\frac{2}{10}=1\frac{1}{5} \qquad 6.\ \frac{\frac{2}{7a}}{b-1}=\frac{2}{7a}\times\frac{1}{b-1}=\frac{2}{7ab-7a}$$

Cross-Multiplying with Fractions:

$$7.\ \frac{y}{9}=\frac{2}{3}, y = \rightarrow 3y=18 \rightarrow y=6$$

$$8.\ \frac{x}{2}=\frac{3x}{y+1}, x\neq 0, y= \rightarrow xy+x=6x \rightarrow xy=5x \rightarrow y=5$$

Note: It's important that *x* cannot be 0 in this problem. As you might know, you can never divide by zero.

Splitting a Fraction:

$$9.\ \frac{2+x}{3}=\frac{2}{3}+\frac{x}{3} \qquad 10.\ \frac{3a-ab}{b}=\frac{3a}{b}-\frac{ab}{b}=\frac{3a}{b}-a$$

Using Fractions to Express a Ratio:

11. The ratio of *x* to *z* is 17:20. $\frac{x}{z}=\frac{17}{20}$

12. The ratio of the length (*l*) of a rectangle to its width (*w*) is 5 to 6. $\frac{l}{w}=\frac{5}{6}$

Drill Explanations | 23

Converting to a Decimal or Percentage:

13. $\frac{3}{5} = 0.6 = 60\%$ 14. $\frac{5}{8} = 0.625 = 62.5\%$ 15. $\frac{4}{3} = 1.3\bar{3} = 133\frac{1}{3}\%$

Percentages

Finding a Percentage *of*:

16. 200% of $16 = 32$ 17. $\frac{1}{2}\%$ **of 50** $= 0.25$

This Is What Percentage of That:

18. 12 is what percentage of 18? $\frac{12}{18} = \frac{2}{3} = 66\frac{2}{3}\%$

19. **2 is what percentage of ½?** $\frac{2}{\frac{1}{2}} = 2 \times 2 = 4 = 400\%$

Percentage Greater Than/Less Than:

20. 12 is what percentage greater than 10? $\frac{2}{10} = 20\%$

21. 15 is what percentage less than 100? $\frac{85}{100} = 85\%$

22. The price of a shirt increased by 5% and then increased by 10%. The final price is what percent more than the original price?

Proceed as if the original price = $100.

$100 + 5\%(100) = 105$. $105 + 10\%(105) = 105 + 10.5 = 115.50$, which is the final price. (For this last step, you can simply multiply 105 by 1.1 to get 115.50. This is 110% of 105, which is what you want.)

The difference between the final price and the original price is 15.50. Therefore, $\frac{15.50}{100} = 15.5\%$.

23. The area of a circle is increased by 20% and then decreased by 20%. The final area is what percent less than the original area?

Proceed as if the original area = 100 square inches, meters, or whatever.

$100 + 20\%(100) = 120$. $120 - 20\%(120) = 120 - 24 = 96$, which is the final area. (You can simply multiply 120 by .80 to get 96. This is 80% of 120, which is what you want. In other words, "20% less than" is the same as "80% of.")

The difference between the final area and the original area is 4. Therefore, $\frac{4}{100} = 4\%$.

Note: If the question had instead asked, "The final area is what percent *of* the original area," then you would place 96 over 100 to get 96%. Be careful with the wording of word problems.

Roots and Exponents

Multiplying with Exponents:

$$24.\ y^3 \times y^8 = y^{11} \qquad 25.\ r^2 r^3 r^4 = r^9$$

Dividing with Exponents:

$$26.\ \frac{x^5}{x^4} = x \qquad 27.\ \frac{x^5}{x^{-2}} = x^{5-(-2)} = x^7$$

Raising a Power to a Power:

$$28.\ (x^5)^3 = x^{15} \qquad 29.\ (y^{-3})^{-10} = y^{30}$$

Negative Exponents:

$$30.\ \left(\frac{1}{2}\right)^{-3} = 2^3 = 8 \qquad 31.\ \frac{ab^{-3}c}{bc^{-3}} = \frac{ac^4}{b^4}$$

Fractional Exponents:

$$32.\ 81^{\frac{1}{2}} = 9 \qquad 33.\ 64^{\frac{2}{3}} = \left(\sqrt[3]{64}\right)^2 = 4^2 = 16$$

Roots:

$$34.\ \sqrt[3]{8} = 2 \qquad 35.\ \sqrt[4]{81} = 3$$

Factoring Roots:

$$36.\ \sqrt{40} = \sqrt{4}\sqrt{10} = 2\sqrt{10} \qquad 37.\ \sqrt{200} = \sqrt{100}\sqrt{2} = 10\sqrt{2}$$

Day 3

Algebra Basics – Simultaneous Equations, Inequalities

After reading this material and completing the drill at the end, sign and check the box for Day 3.

Algebra

Anytime you're solving an equation that has an unknown value, usually represented by a letter such as x, you're doing algebra. This might seem obvious, but never forget that x, y, or whatever is **just a number**. It has no magical properties. If you're comfortable with the number 7, you should be just as comfortable with x. The same thing goes for xy or $\frac{a-9}{2b}$. This might not look like "a number," but it is in the same way that $\frac{10-2}{2(4)}$ is "a number"—the number 1.

Algebra can be tested on the SAT with just a bunch of known and unknown values (numbers and letters), or it can represent some real-world situation (a word problem).

Solving for an unknown: A typical algebra question gives you an equation and asks you to **solve for the unknown**. To do this, you'll get the unknown on one side of the equation—isolating it—by adding, subtracting, multiplying, dividing, raising to a power, or finding a root to **both sides.**

Simple: $2x - 9 = x$
By adding 9 to both sides and subtracting x from both sides, we get $x = 9$.

Challenging: $\frac{y+3}{7} = -\frac{y}{2}$
Here, it's a good idea to cross-multiply.
That gives us $2(y + 3) = -7y$.
We now need to distribute the 2, giving us $2y + 6 = -7y$.
We now *collect like terms* by adding 7y to both sides and then,
while we're at it, subtracting 6 from both sides.
That gives us $9y = -6$.
Our final step is to divide each side by 9 and simplify.
$y = -\frac{6}{9} = -\frac{2}{3}.$

Note: Cross-multiplication really is just a shortcut to "do the same thing to both sides." You could have started out instead by multiplying each side by 7 and then multiplying each side by 2. It works out the

26 |Day 3

same. Also, the fraction $-\frac{y}{2}$ can be thought of as negative y over 2 *or* as y over negative 2, but *not* negative y over negative 2.

What goes wrong/what takes too long? Anytime you need to take several steps to solve for an unknown (you usually will), careless errors can happen. One way to avoid this is to *not* think "I'm moving x over to the other side of the equation." You're not *moving* anything. You're adding, multiplying—whatever—to **both sides of the equation.** If you keep this in mind, you're less likely, for example, to subtract 3 from only 1 side.

Solving for a combination of unknowns: You might have one equation with more than 1 unknown value. You can't solve for both with only 1 equation, but you might need to combine them in some way.

Simple: If $r = 3 - s$, what is the value of $r + s$?
This is a good example of how important it is to focus on what the question is looking for.
To solve, just add s to both sides: $r + s = 3$.

Challenging: If $\frac{6}{5a} = b$, what is the value of ab?
As with many equations involving fractions, it's a good idea to "flatten the equation" by multiplying each side by $5a$. This gives us $6 = 5ab$. Now we just need to divide each side by 5, giving us $\frac{6}{5} = ab$.

What goes wrong/what takes too long? You must pay attention to what the questions is asking. If, for example, you lose sight of the fact that the question is asking you to solve $\frac{r}{x}$ and you try to solve for x, you're in trouble. **Your very last step before choosing an answer** is to ask yourself, "What exactly did they want?" Which brings us to the following.

Solving for a combination of unknowns and knowns: This is similar to the above in that you're not simply solving for, let's say, x.

Simple: If $3x = 5$, what is $2x$?
Step 1: Divide each side by 3 to isolate x: $x = \frac{5}{3}$.
Step 2: **Do not** choose $\frac{5}{3}$ as the answer since the question did not ask for the value of x.
And yes, there's a good chance that $\frac{5}{3}$ will be among the choices.
Instead, multiply each side by 2: $2x = \frac{10}{3}$. Perhaps they want the answer expressed as $3\frac{1}{3}$.

Note: Remember, some of the questions will ask you to type in the correct answer rather than selecting it from 4 choices. In the question above, you **cannot** type $3\frac{1}{3}$ since it will look like 31/3, which is incorrect. You should type 10/3 or—since $3\frac{1}{3}$ is equal to $3.\overline{3}$—you can type 3.333 in the space provided. (There's no way to enter the repeating bar symbol, so you'd have to enter as many numbers as you can in order to approximate. You are given 5 "spaces" to enter a number, or 6 spaces if the number requires a negative symbol.)

Challenging: If $a = 11 + 3b$, solve for $2a - 6b$.
Step 1: Subtract $3b$ from each side: $a - 3b = 11$.

Step 2: Double each side.
In other words, multiply each side by 2: $2a - 6b = 22$.
You're done.

What goes wrong/what takes too long? Do not fall into the trap of always solving for x or some other unknown value. You might need to solve for $2x$ or something else. Also, don't **oversolve**. If you lose sight of what the question is asking, you might go right past it trying to solve for something else. For example, if they ask you to solve for $2x - 3$ and you get there but keep going until you get x solved, that would be a time-wasting shame.

Solving for an unknown in terms of another: In this case, you're still isolating an unknown, but the other side of the equation will also have 1 or more unknowns.

Simple: $5 - x = z$, then $x = ?$
To isolate x, you can add it to both sides and subtract z from both sides.
(Note: We're not *moving things*; we're adding to and subtracting from both sides.)
This gives us $5 - z = x$. We have now *solved for x in terms of z*.

Challenging: If $\frac{3-a}{b} = 2a$, what is a in terms of b?
Step 1: Multiply each side by b (flattening the equation) $3 - a = 2ab$
Step 2: Add a to each side (collecting your unknowns) $3 = 2ab - a$
Step 3: This is an important step and one that students often forget.
We're going to *factor* an a from the 2 terms on the right (more about this later).
This gives us $3 = a(2b - 1)$. Now we simply divide each side by $2b - 1$.
This gives us $\frac{3}{2b-1} = a$.
We're done. We gave them a in terms of b.

What goes wrong/what takes too long? Once again, you must focus on exactly what the question is asking. But now let's talk about the unappreciated art of factoring.

Factoring: This might appear to violate the rule that you always do the same thing to both sides of an equation. But it doesn't because you're not changing the value of 1 side; you're just changing how it **looks.** You're expressing it differently. You're breaking it into the product of 2 (or more) things, the same way that 12 is equal to 4 times 3.

Let's say you have the expression $xy + 2x$. Note that this is **not** an equation; it's just an expression. We can factor (divide out) an x from each of these 2 terms, giving us $x(y + 2)$. We haven't changed the value of the expression at all, just its **look**. After all, if we were instead given $x(y + 2)$, we could have gone in the opposite direction by **distributing** the x, giving us $xy + 2x$. Factoring and distributing are the opposites of each other.

So why factor? We saw an example in the challenging question above. Let's try another.

Simple: If $xy + 2x = 50$, solve for x.
Step 1: Factor an x from the 2 terms on the left: $x(y + 2)$.
We now have $x(y + 2) = 50$.
Step 2: Divide each side of the equation by $(y + 2)$: $x = \frac{50}{y+2}$

Done.

Challenging: If $r + 2s = st - 4$, what is s in terms of r?
Step 1: Collect the unknowns by subtracting st from both sides.
At the same time, let's subtract r from both sides: $2s - st = -4 - r$.
Step 2: Factor s from the 2 terms on the left: $s(2 - t) = -4 - r$.
Step 3: Divide each side by $2 - t$: $s = \frac{-4-r}{2-t}$.
Step 4: You've done the work correctly, but let's say
you don't see your answer among the choices.
Step 5: Multiply the top and bottom of the right side
of the equation by negative 1.
This doesn't change the value of anything
because multiplying by $\frac{-1}{-1}$ is the same as multiplying by 1.
But once again, it changes its **look**, giving us $s = \frac{4+r}{-2+t}$,
which is still not the way it looks in the choices.
But how about $\frac{r+4}{t-2}$? It's the same thing, right?
Yeah, that's probably how they would want it to look.

What goes wrong/what takes too long? We could have avoided some of that work by using a different Step 1. We could have added 4 and subtracted $2s$ from each side. But if you don't take the shortest route to the answer, you might need to do some re-expressing as we did in Step 5. The biggest problem with factoring, though, is that students simply forget that they can do it, and they get stuck. So whenever you're stuck, **consider factoring**.

Solving with Multiple Equations: This is also known as *simultaneous equations,* and it's a biggie. These **will** show up on the test. If you need to solve for 2 unknowns, you will need 2 equations; for 3 unknowns, you'll need 3 equations, and so on. There are 2 main ways to solve these: **substitution** and **elimination**.

Simple: $3x - y = 2$ and $x + y = 6$.
Solve for x and y. First, we'll solve using substitution.
We begin by getting 1 unknown in terms of the other.
For example, we could take the 2nd equation and subtract y from both sides.
This gives us $x = 6 - y$.
From here on, anytime you see x, you'll replace it with $6 - y$,
which is, after all, the same number.
That's substitution.
So the first equation is now $3(6 - y) - y = 2$.
This is what we almost always want: **1** equation with only **1** unknown.
Now we can easily solve it.

Step 1: Distribute the 3, giving us $18 - 3y - y = 2$.
Step 2: Collect like terms, giving us $18 - 4y = 2$
and then $-4y = -16$ or simply $4y = 16$.
Step 3: Divide each side by 4, giving us $y = 4$. That's solved!
Step 4: Now that we know that $y = 4$,

we can use this information—with either equation—to solve for x.
The second equation is simply $x + y = 6$, so we now know that $x = 2$.

Let's try the other major approach to solving simultaneous equations — **elimination**.

Here's the same question: Given $3x - y = 2$ and $x + y = 6$, solve for x and y. We'll now add the equations to **each other** (in some cases, we'll subtract them from each other). Some students think of this as *stacking* since it can be done simply by putting 1 equation over the other and then either adding or subtracting.) Here, we'll add.

$$3x - y = 2$$
$$+ x + y = 6$$
$$\overline{}$$
$$4x = 8$$

In 1 simple step, we got **1** equation with **1** unknown. That's what we (usually) want. We can now obviously divide each side by 4 to get x and then use that information in either equation to get $y = 2$.

Elimination can be a time-saver, as we see here, but it doesn't always play out so nicely. Let's change the question above just a little bit.

$3x - y = 2$ and $x + 2y = 6$. Solve for x and y.
Note: If we added the 2 equations, nothing drops out.
We won't have **1** equation with **1** unknown.
So you might decide to use substitution on this one.
Or you could multiply every term in the 1st equation by 2
and **then use elimination** since now, by adding the equations together:

$$6x - 2y = 4$$
$$+ x + 2y = 6$$
$$\overline{}$$
$$7x = 10$$

The y-values dropped out, so now we can solve for x and then use that info to solve for y. We won't bother to do that here. But let's look at this one:

$2a + 5b = 13$ and $a + 5b = 14$.
Solve for b. If we use elimination here, we'll have
$$2a + 5b = 13$$
$$a + 5b = 14$$
$$\overline{}$$

Be careful. If we add these together, nothing gets eliminated. What we can do is subtract one from the other. (You could also think of this as multiplying every term in 1 of the equations and then **adding** them. It's the same thing.) That leaves us with $a = -1$. Is this the answer? No, because the **question** asked for b. But it's a simple matter to use the fact that $a = -1$ to get b, which is 3.

Challenging: $6r + 7s = 16$ and $-r + 2s = 10$. Solve for r.

It's usually a good idea to first consider elimination since it's so efficient. Adding or subtracting these 2 together won't eliminate anything, but you could multiply every term in the 2nd equation by 6 and then add. That leaves us with only $19s = 76$, which means that $s = 4$.

Be careful because this might be 1 of the choices but not the right one. We want r, so if you now replace s with 4 in either equation, you'll get the correct choice: -2.

You should be able to use substitution here quite efficiently too. The more tools you have, the better. In the 2nd equation, we could add r to both sides and subtract 10 from both sides. That leaves us with $r = 2s - 10$. We can now use that information in the 1st equation, giving us $6(2s - 10) + 7s = 16$. With our **1** equation with **1** unknown, we can solve the usual way and then find the value for r.

What goes wrong/what takes too long? Anytime you have a multi-step process, careless errors can creep in at any step. With these simultaneous equations, make sure with elimination that you either consistently add or subtract each term of the equations. If you're subtracting, remember that subtracting a negative is the same as adding. If you're distributing at any step, make sure you distribute completely, multiplying by every term within the parentheses.

As for shortcuts, here's a nice one. Let's say you are given the following question: $4a - 3b = 39$ and $9a + 19b = 165$. The question says to solve for $13a + 16b$. That looks like it could eat up a lot of time whether you use substitution or elimination. But remember, it's very important that you pay attention to what the question really wants and to remember that long calculations are rarely involved in SAT questions. Did you see the shortcut? If you simply add the 2 equations together without doing anything to them, you'll get $13a + 16b = 204$. Done!

Now some of you might be thinking, "What about that **on-screen calculator?** That will solve these simultaneous equations fast and easy." Well, you're right, and we'll discuss that in depth soon. But you should know how to do it algebraically. That will pay off.

Inequalities: These use the signs for "greater than" (>) and "less than" (<) to solve **not** for a missing value but for a **range** of values. However, we handle them almost (that's an important **almost**) exactly the same way we handle regular equations, doing something to each side in order to isolate the unknowns.

Simple: $2x - 5 > 9$.
Step 1: Add 5 to each side, giving us $2x > 14$.
Step 2: Divide each side by 2, giving us $x > 7$.
There is an infinite number of possibilities or x,
but we do know that it's greater than 7.
And that's all we know. It could, for example, be 7.01.

Now let's focus on the word *almost* mentioned above. As soon as you see an inequality on the test, think this: "The SAT people want to know if **I know** that the inequality sign changes direction when **each side is multiplied or divided by a negative value**."

So let's look at $5 - 3x < 20$.
Step 1: Subtract 5 from each side, giving us $-3x < 15$.

Step 2: Divide each side by negative 3, and **flip the sign**.
This gives us $x > -5$.

Challenging: $-7 < 4x + 1 < 33$
This is a *compound* inequality, meaning that there are more than 2 signs.
The best way to read these is from the center outward:
"$x + 1$ is a value between -7 and 33."
Step 1: Subtract 1 from each "side." $-8 < 4x < 32$
Step 2: Divide each term by 4. $-2 < x < 8$

Note: We did not need to flip the sign because we did not multiply or divide by a negative value.

What goes wrong/what takes too long? Forgetting to flip the sign is a common error. And it's worth keeping the number line in mind with inequalities. A few questions ago, we found that $x > -5$. This means that x is to the right of negative 5 on the number line. It might be -4.2, for example.

*

We'll soon get to quadratic equations and word problems, but first—a drill. These questions will help you solidify your knowledge of the above material as well as some of the material from Day 2. If you're **very** confident about this material, you can do only the **bold-faced** questions. And remember to be organized in your work. Just write what needs to be written.

Day 3

Drill

Solving for an Unknown: Get the unknown on 1 side of the equation—isolating it—by adding, subtracting, multiplying, dividing, raising to a power, finding a root to **both sides**.

Solve for the unknown value(s). And be **efficient!**

1. $2 + 3s = -1$
2. $\frac{9+z}{10} = \frac{z-1}{4}$

Solving for a Combination of Unknowns: You can't solve for both with only 1 equation, but you might need to combine them in some way.

3. $a = 17 - b, a + b =$
4. $2x = 7y, \frac{x}{y} =$

Solving for a Combination of Unknowns and Knowns: This is similar to the above in that you're not simply solving for, let's say, x.

5. $5z = z + 9, -4z =$
6. $\frac{3}{t+1} = \frac{5}{t}, 4t =$

Solving for 1 Unknown in Terms of Another: In this case, you're still isolating an unknown, but the other side of the equation will also have 1 or more unknowns.

7. $a = 3b + 2, b =$
8. $\frac{3}{x} = \frac{5y-1}{2y}, 10xy =$

Factoring: You're not changing the value of 1 side; you're just changing how it **looks**.

9. $5y + xy = -6, y =$
10. $a - ab = \frac{2}{b}, a =$

Solving with Multiple Equations: There are 2 main ways to solve these: substitution and elimination.

11. $2y - x = 16$ and $5y + x = 54, y =$

12. $a + 2b = 5$ and $4a - 3b = -13, b =$

13. $3a + 11b = -52$ and $-a + 2b, 4a + 9b =$

Inequalities: In these questions, isolate the unknown, solving for its range of possible values.

14. $a - 12 < 2a$ 15. $\dfrac{x}{-5} < x + 1$ **and** $x > 0$

BONUS QUESTIONS

These questions combine material from Days 2 and 3. Bonus does not mean "optional." It means "extra helpful"!

16. x is 5% of y and $y - x = 304$, $x =$

17. $\dfrac{\frac{3}{a}}{\frac{5}{b}} = 60$, $\dfrac{b}{a} =$

18. $\dfrac{1-a}{12} + \dfrac{7}{b} = 2$, $2 - \dfrac{b-ab}{12b} =$

19. $\dfrac{\left(\frac{1}{2}\right)^3}{x} > 8$ and $x > 0$

20. $\sqrt{128} = 8t$ $t =$

21. The ratio of x to y is 4:9, and the ratio of x to z is 16:1. What is the ratio of z to y?

22. 250 is 300% more than a, and a is 200% of b. What is the value of b?

Day 3

Drill Explanations

Solving for an Unknown:

$$1.\ 2 + 3s = -1$$
Isolate s: Subtract 2 from both sides and divide by -3.
□ $3 = -3s \quad s = -1$

$$2.\ \frac{9+z}{10} = \frac{z-1}{4}$$
Cross-multiply, distribute, and isolate z.
□ $36 + 4z = 10z - 10 \quad 6z = 46 \quad z = \frac{46}{6} = 7\frac{4}{6} = 7\frac{2}{3}$

Solving for a Combination of Unknowns:

$$3.\ a = 17 - b, a + b =$$
Isolate $a + b$: Add b to both sides.
□ $a + b = 17$

$$4.\ 2x = 7y, \frac{x}{y} =$$
To get this in 1 step, divide each side by $2y$.
That way you'll get y under x without the 2 on top.
That's what you want.
You could also do it in 2 steps, first dividing by y and then by 2, like this:
□ $\frac{2x}{y} = 7 \quad \frac{x}{y} = \frac{7}{2} = 3\frac{1}{2}$

Solving for a Combination of Unknowns and Knowns:

$5.\ 5z = z + 9, -4z =$ Isolate $-4z$ by subtracting $5z$ and 9 from both sides.
□ $\quad -4z = -9$

$6.\ \frac{3}{t+1} = \frac{5}{t}, 4t =$ Isolate $4t$, starting with cross-multiplication.
□ $\quad 5t + 5 = 3t \quad 2t = -5 \quad 4t = -10$

Solving for 1 Unknown in Terms of Another:

$7.\ a = 3b + 2, b =$ To isolate b, subtract 2 and divide by 3.

☐ $a - 2 = 3b$ $\frac{a-2}{3} = b$

8. $\frac{3}{x} = \frac{5y-1}{2y}, x =$ To isolate x, cross-multiply, **factor** out the x, and divide.

$$6y = 5xy - x \quad 6y = x(5y - 1) \quad x = \frac{6y}{5y - 1}$$

Factoring:

9. $5y + xy = -6, y =$ To isolate y, **factor** it out and divide by $(5 + x)$.
☐ $y(5 + x) = -6 \quad y = -\frac{6}{5+x}$

10. $a - ab = \frac{2}{b}, a =$ To isolate a, **factor** it out and divide by $1 - b$.
☐ $a(1 - b) = \frac{2}{b} \quad a = \frac{2}{b(1-b)} \quad a = \frac{2}{b-b^2}$

Solving with Multiple Equations:

11. $2y - x = 16$ and $5y + x = 54, y =$
This can be simply solved with elimination.
Add the 2 equations to each other, and the x values are eliminated.
☐ $7y = 70 \quad y = 10$

12. $a + 2b = 5$ and $4a - 3b = -13, b =$
For this, we'll use substitution.
We subtract $2b$ from each side of the 1st equation to isolate a.
We then substitute this into the 2nd equation and solve.
(If you used elimination efficiently, great!
And yes, the on-screen calculator can be used, as we'll discuss soon—promise!)
☐ $a = 5 - 2b \quad 4(5 - 2b) - 3b = -13 \quad 20 - 8b - 3b = -13$
$20 - 11b = -13 \quad 33 = 11b \quad b = 3$

13. $3a + 11b = -52$ **and** $-a + 2b = 12, 4a + 9b =$
Let's see what happens when we eliminate by subtracting
the 2nd equation from the 1st.
☐ $3a - (-a) = 4b$ and $11b - 2b = 9b$ and $-52 - 12 = -64$
☐ That means that $4b + 9b = -64$.

Inequalities:

14. $a - 12 < 2a$ All you need to do is subtract a from both sides.
☐ $a > -12$

15. $\frac{x}{-5} < x + 1$ **and** $x > 0$.
We can multiply each side by negative 5, but we must flip the sign direction.
☐ $x > -5x - 5 \quad 6x > -5 \quad x > -\frac{5}{6}$

36 |Day 3

BONUS QUESTIONS

16. x is 5% of y and $y - x = 304$, $x =$

We can write the first equation as $x = \frac{1}{20}y$. (If you want to use 0.05 instead, that's fine.)
Now that we've isolated x, we can use substitution,
giving us $y - \frac{1}{20}y = 304$. This simplifies to $\frac{19}{20}y = 304$.
Multiplying each side by $\frac{20}{19}$ gives us $y = 320$.
The question asks for x, and we can use either equation to do that.
$320 - x = 304$, so $x = 16$.

17. $\frac{\frac{3}{a}}{\frac{5}{b}} = 60, \frac{b}{a} =$ First simplify the fraction and rewrite the equation.

- $\frac{\frac{3}{a}}{\frac{5}{b}} = \frac{3}{a} \times \frac{b}{5} = \frac{3b}{5a} \quad \frac{3b}{5a} = 60$.

- Then divide each side by $\frac{3}{5}$, giving you $\frac{b}{a} = 100$

18. $\frac{1-a}{12} + \frac{7}{b} = 2, 23b =$

- First, add the 2 fractions with a common denominator. $\frac{b - ab + 84}{12b} = 2$
- Now cross-multiply: $24b = b - ab + 84$
- Subtracting b from each side gives you $23b = -ab + 84$ or $23b = 84 - ab$

19. $\frac{\left(\frac{1}{2}\right)^3}{x} > 8$ and $x > 0$

Since x is positive, we multiply and keep the sign direction.
$\left(\frac{1}{2}\right)^3 > 8x$. This simplifies to $\frac{1}{8} > 8x$ and then,
dividing by 8, we get $x < \frac{1}{64}$.

20. $\sqrt{128} = 8t \quad t =$
We can factor $\sqrt{128}$ into $8\sqrt{2}$.
The equation is now $8\sqrt{2} = 8t$, meaning that $t = \sqrt{2}$

21. The ratio of x to y is 4:9, and the ratio of x to z is 16:1.
What is the ratio of z to y? First we'll make the ratios into fractions.

$$\frac{x}{y} = \frac{4}{9} \text{ and } \frac{x}{z} = \frac{16}{1}$$

One way to solve this is to multiply the top and bottom of the
right-hand side of the 1st equation by 4, giving us $\frac{x}{y} = \frac{16}{36}$.
Now that we have 16 on top of both equations,
we can say that the ratio of z to y is 1:36.

22. 250 is 300% more than a, and a is 200% of b.

Drill Explanations | 37

What is the value of b?

Since 250 is 300% **more than** a (not 300% **of** a),
we can say that $250 = 4a$.
(If 250 is 100% **more than** a, it would be twice a.)
Dividing each side by 4 gives us $a = 62.5$.
On the other hand, since a is 200% **of** b, we can write $a = 2b$.
Since $a = 62.5$, we know that $62.5 = 2b$,
and dividing each side by 2 gives us $b = 31.25$.

Day 4

Quadratic Equations: Grouping, Completing the Square, Quadratic Formula, OSC

After reading this material and completing the drill at the end, sign and check the box for Day 4.

So far, we've looked mostly at *linear* algebraic equations where no unknown value is squared or raised to some other power. A linear equation, when graphed, creates a line. (We'll look at this in more depth later in the course.) But the SAT loves to present you with *quadratic* equations where unknown values are squared. These equations, when graphed, create parabolas. Today, we'll explore solving these equations, as well as equations involving higher powers such as when a term is cubed.

A quadratic **expression** (not an *equation* because there's no equal sign) is often given in the form $ax^2 + bx + c$. For example, $5x^2 + 2x + 9$ is a quadratic expression. On the SAT, you'll often see quadratic expressions where $a = 1$. Since there's no reason to write the 1, they'll look, for example, like $x^2 + 7x + 12$.

Let's look at a type of equation that can be thought of as a building block of a quadratic equation: $x^2 = 64$.

This is a simple equation, but you **must** be aware that it has 2 solutions: 8 and negative 8. Forgetting about the negative 8 is a good way to fall into an SAT trap. Many (not all) quadratic equations have 2 solutions.

Now let's look at one of the actual quadratic expressions mentioned above: $x^2 + 7x + 12$. One thing we can do is **re-express** or re-form it. Being able to re-express algebraic expressions is a key skill. We've already talked about **factoring**—breaking up a number into 2 or more parts. When you multiply the parts, you're back where you started. With a quadratic such as this one, we want to factor it into 2 parts that each has 2 terms. Each one is called a binomial (2 terms). It'll look like this: $(x\ )(x\ )$. For the missing values, we want 2 numbers whose product is 12 and whose sum is 7. Those 2 numbers are 3 and 4, so we can say that $x^2 + 7x + 12 = (x + 3)(x + 4)$. You might know the term FOIL, which is the opposite operation. If we take $(x + 3)(x + 4)$ and distribute (multiply) the values in the following order—**F**irst terms, **O**uter terms, **I**nner terms, and **L**ast terms—we're back to $x^2 + 7x + 12$.

Why factor? Some of you might already be thinking about solving for x, but we can't because we don't have an equation. But okay, let's set our expression equal to something, and yes, let's set it equal to 0, which is often the case on the exam.

Solving Simple Quadratic Equations Set to Zero: Factor the Quadratic into 2 Binomials, and Solve

If you see $x^2 + 7x + 12 = 0$, once you factor, you have $(x + 3)(x + 4) = 0$. Now you can solve. We know that either $(x + 3)$ or $(x + 4)$ must be equal to 0. There's no other way the product of 2 numbers can equal 0. **Here's an important note.** We're saying that $(x + 3)$ must equal 0 or that $(x + 4)$ must equal 0; we are **not** saying that x equals 0. But we're almost done. We have $x + 3 = 0$ and $x + 4 = 0$. That means our 2 solutions are $x = -3$ and $x = -4$. Let's try that some more.

Simple: $x^2 + 11x + 28 = 0$
We can factor this into $(x + 4)(x + 7) = 0$.
The solutions are $x = -4, x = -7$

Challenging: $x^2 + 7x - 60 = 0$
We can factor this into $(x - 5)(x + 12) = 0$.
The solutions are $x = 5, x = -12$

If it seems like too much hit and miss, we'll discuss other methods of factoring, but for this equation, it's probably best just to think about all the factor pairs with a product of negative 60. One has to be negative, and one must be positive. Some can be eliminated quickly, such as 2 and 30, which will clearly not sum to positive 7, regardless of which of these numbers is negative.

Solving Simple Quadratic Equations *Not* Set to 0: If Possible, M*ake* It Equal to 0

What if our quadratic expression is not equal to 0? Well, let's say you're given $x^2 + 9x + 5 = -15$. Since we do typically solve these equations by "setting them equal to 0," let's add 15 to both sides. That gives us $x^2 + 9x + 20 = 0$. And **now** we can factor. The 2 values whose product is 20 and whose sum is 9 are 4 and 5. Therefore, $x^2 + 9x + 20 = 0$ becomes $(x + 4)(x + 5) = 0$, and the solutions are $x = -4$ and $x = -5$.

Simple: $x^2 + 10x = -16$
Let's add 16 to both sides so we get $x^2 + 10x + 16 = 0$.
Now we can factor this into $(x + 8)(x + 2) = 0$,
and the solutions are $x = -8$ and $x = -2$.

Challenging: $x^2 - x = 56$
Here we'll subtract 56 from both sides, giving us $x^2 - x - 56 = 0$.
Now we can factor this into $(x - 8)(x + 7) = 0$,
and the solutions are $x = 8$ and $x = -7$.

Signs (+ and -) can become confused when solving quadratic equations, so **be careful** not to get it backward. Also, if you see an equation such as the challenging question above, remember that there is an "invisible" negative 1 in front of the middle term. (Those numbers in front of an unknown value like x are called *coefficients*.)

Factoring Quadratics without a Middle Term: The Difference of 2 Squares

One quadratic expression mentioned above was $x^2 - 9$. This 2-term quadratic expression has its own name: the difference of 2 squares. In other words, it's x squared minus 3 squared. We factor this into $(x - 3)(x + 3)$. If you use FOIL to distribute the terms, you'll see that it does, indeed, give us $x^2 - 9$.

Simple: $x^2 - 49$ This factors into $(x - 7)(x + 7)$.

Challenging: $y^2 - \frac{1}{4}$ This factors into $(y - \frac{1}{2})(y + \frac{1}{2})$.

Quadratics Equations with 1 Solution: Look for a Perfect Square in the Last Term (the *Constant*)

We mentioned above that some quadratic equations have only 1 solution. For example, let's say you have $x^2 + 10x + 25 = 0$. What 2 numbers give us a product of 25 and a sum of 10? Well, 5 and 5. Therefore, we can write this in factored form as $(x + 5)(x + 5) = 0$ or simply $(x + 5)^2 = 0$. The one and only solution is -5. This sometimes happens when the final term, called the *constant*, is a perfect square, like 25.

Simple: $x^2 + 8x + 16 = 0$.
We can write this in factored form as $(x + 4)(x + 4) = 0$
or simply $(x + 4)^2 = 0$.
The one and only solution is $x = -4$.

Challenging: $x^2 - 22x + 121 = 0$.
We can write this as $(x - 11)^2 = 0$.
The solution is $x = 11$.

Solving Quadratics through Grouping: Re-express the Quadratic as 4 Terms, and Then Group Them into 2 Pairs

So far, we've only looked at quadratics that could be factored just by playing around with some numbers in our head. This often works, but when the coefficient of the x^2 term is greater than 1, as in the example $6x^2 + 5x - 6$, the factored form of this expression is not so obvious.

Let's say you have the equation $6x^2 + 5x - 6 = 0$.
One way to factor this quadratic expression is through a process called *grouping*.

Step 1: Find the product of the coefficient of the x^2 term — 6
and the last term, negative 6. We multiply 6 by negative 6 to get negative 36.
Step 2: Find the 2 numbers whose sum is the coefficient of the
middle term (as we usually do) and whose product is
the number we got in step 1.
Those numbers are **negative 4 and 9**.
In other words, $-4 + 9 = 5$ and $(-4)(9) = -36$.
Step 3: Rewrite the original quadratic as $6x^2 - 4x + 9x - 6$.
It's the same expression, the same **information** as the original
but re-expressed as 4 terms.
Step 4: This is where we use parentheses to **group**.
We'll group the first 2 and then the second 2 terms,
giving us $(6x^2 - 4x) + (9x - 6)$.
Step 5: Now we factor the largest value we can out of each of the 2 groups.
For the first group, we can factor out $2x$.
That gives us $2x(3x - 2)$.
For the second group, we can factor out 3.
That gives us $3(3x - 2)$.
You'll note that both of these factored groups have $(3x - 2)$.
That's great. Because now—
Step 6: We can think of the entire quadratic as $2x(3x - 2) + 3(3x - 2)$,
and since we have that $3x - 2$ in both terms,
we can factor it out, giving us $(3x - 2)(2x + 3)$.
That's the factored form of our original quadratic.
Step 7: Since $(3x - 2)(2x + 3) = 0$, we can say that $3x - 2 = 0$ and $2x + 3 = 0$.
These 2 simple equations can be easily solved, giving us $x = \frac{2}{3}$ and $x = -\frac{3}{2}$.

The above explanation of grouping might seem lengthy and intricate. It does take some getting used to, but it takes much longer to explain than to do. We'll look at a problem here and more later in the drill after the lesson.

Simple/Challenging: $4x^2 + 8x - 5 = 0$
We multiply the 4 and negative 5 to get negative 20.
We then need 2 numbers whose product is negative 20 and whose sum is 8.
These 2 numbers are negative 2 and positive 10.
We then rewrite the equation with 4 terms: $4x^2 - 2x + 10x - 5 = 0$.
(Tip: It's usually better to have the positive number—positive 10 in this case—be
the **3rd** term, not the 2nd.)
Now we can form our 2 groups using parentheses. $(4x^2 - 2x) + (10x - 5) = 0$.
Next, we factor the largest value we can out of each of these 2 groups.
That gives us $2x(2x - 1) + 5(2x - 1) = 0$.

Since we have $2x - 1$ in each group, we can factor it out,
giving us $(2x - 1)(2x + 5) = 0$.
Therefore, the solutions are $x = \frac{1}{2}$ and $x = -\frac{5}{2}$.

Solving by Completing the Square: Express a Quadratic in the Vertex Form $a(x - h)^2 + k$

Sometimes you'll need to re-express a quadratic so it no longer looks, for example, like $x^2 + 6x - 10$. Instead, we want it to look like $(x + 3)^2 - 9$. (If you compare this to the bold-faced form given above, the a value is 1, so there's no need to write it.) To do this, we'll *complete the square*. This takes a few steps.

Step 1: Starting with $x^2 + 6x - 10$, we'll take **half** of the coefficient
of the middle term (half of 6 is 3) and write $(x + 3)^2$.
Step 2: If we multiply this out, we would get $(x + 3)(x + 3) = x^2 + 6x + \mathbf{9}$.
Since this is not what we started with, we need to **subtract 19.**
This would now be equivalent to $x^2 + 6x - 10$.
Step 3: We write $(x + 3)^2 - 19$.

Why do this? Why complete the square? This form of a quadratic is called the **vertex form**. We'll see a good use of this way of expressing a quadratic once we get into word problems and coordinate geometry.

Simple: Rewrite $x^2 + 8x + 13$ by completing the square.
We take half of the coefficient of 8 and write $(x + 4)^2$.
When multiplied out, this is $x^2 + 8x + \mathbf{16}$.
Since this is 3 more than the original, we must subtract 3, giving us $(x + 4)^2 - 3$.
This is the same *information* as the original but in different form.

Challenging: Rewrite $y^2 + y - \frac{3}{4}$ by completing the square.
We take half of the coefficient of 1 and write $(y + \frac{1}{2})^2$.
When multiplied out, this is $y^2 + y + \frac{1}{4}$.
Since this is 1 more than the original
(the difference between $-\frac{3}{4}$ and $\frac{1}{4}$ is 1),
we must subtract 1, giving us $(y + \frac{1}{2})^2 - 1$.

Solving by the Quadratic Formula: $x = \frac{-b \pm \sqrt{b^2 - 4ac}}{2a}$

This works every time. No matter how odd the quadratic looks, this formula will give you the solution. It might look scary, but you'll get used to it. You might never need it on the test, but if you do—if there's no other efficient way to solve for x—if you **know** this formula and can **use** this formula, you'll be ahead of the pack.

Note: When we solve for *x* using this equation, we're solving for the solution(s) that will give us 0. These solutions are sometimes called *roots*.

Remember from the beginning of today's lesson that quadratic expressions are often in the form $ax^2 + bx + c$, like $x^2 + 9x + 18$. In this quadratic, *a* is 1. *b* is 9, and *c* is 18. Now we just plug these values into the formula:

$$x = \frac{-9 \pm \sqrt{9^2 - 4(1)(18)}}{2(1)}$$

This simplifies to $x = \frac{-9 \pm \sqrt{81-72}}{2}$, which becomes $x = \frac{-9 \pm \sqrt{9}}{2}$, and then $\frac{-9 \pm 3}{2}$.
We then get $\frac{-6}{2}$, which equals -3, as well as $\frac{-12}{2}$, which equals -6.
Those are the 2 solutions or *roots* that will make the quadratic expression equal 0.

If that seemed unnecessarily complicated, you're right. This quadratic can be easily factored into $(x + 3)(x + 6)$, giving you the same answers without much trouble once you set this equal to 0. However, let's try an equation that doesn't factor so simply.

Simple/Challenging: $x^2 + 9x + 3$
In this case, *a* is again 1, $b = 9$, and $c = 3$.
Now we just plug in these values to get $\frac{-9 \pm \sqrt{9^2 - 4(1)(3)}}{2(1)}$.
This simplifies to $\frac{-9 \pm \sqrt{81-12}}{2}$, and then $\frac{-9 \pm \sqrt{69}}{2}$.
This gives us 2 numbers, $\frac{-9+\sqrt{69}}{2}$ and $\frac{-9-\sqrt{69}}{2}$.
And that's it.
Since the square root of 69 is an irrational number,
there's no way to simplify these choices any further.
These are the 2 solutions or roots of the quadratic.

It might help you remember that *a* and *b* show up twice in the formula, but *c* only shows up once. It's also worth pointing out that the part of the formula that is under the square root sign, $b^2 - 4ac$, has its own name and purpose. We call this part the *discriminant*. Here's why it deserves its own name.

- ☐ If the discriminant is positive, there will be 2 solutions. (In our example above, 69 is positive, so it had 2 solutions.)
- ☐ If the discriminant is 0, there will be only 1 solution. (If you try $x^2 + 10x + 25$, our example above of a quadratic with only 1 solution, you'll see that this is true.)
- ☐ If the discriminant is negative, there will be no *real* solution. (If you're thinking, so what *does* it have, you might not be completely familiar with *imaginary* numbers. Imaginary numbers do have a purpose, but as it now stands, they will not appear on the test.)

There's one more thing. We've only looked at quadratics that have a positive coefficient to the squared value. However, we'll also run into negative coefficients in quadratics such as $-3x^2 - 5x + 2$. We'll discuss these more when we get into coordinate geometry, but here's a sneak peek (which you might already know about). As mentioned at the start of today's lesson, quadratics typically can be graphed as parabolas. If the coefficient of the 1st term is positive, the parabola opens upward. If that term is negative, it opens downward.

Solving with the On-Screen Calculator (OSC): Enter the Quadratic with *x* and *y*

(Some of you were no doubt waiting for this.)

The proper use of the OSC can sometimes be an efficient way to handle quadratics. We'll look at this in greater detail on another day, but here are the basics. Unlike what is almost certainly true on your handheld calculator, the OSC allows you to enter quadratics simply and then see the corresponding **parabola** formed by graphing a quadratic. By zooming in or out, you should then be able to see the vertex of the parabola and the 0's (any points where the parabola crosses the *x*-axis.) Let's look at 1 of the above examples to see how we could use the OSC to solve it.

We solved the equation $6x^2 + 5x - 6 = 0$ through grouping. If, instead, we enter the quadratic $6x^2 + 5x - 6$ into the OSC, we get the following graph.

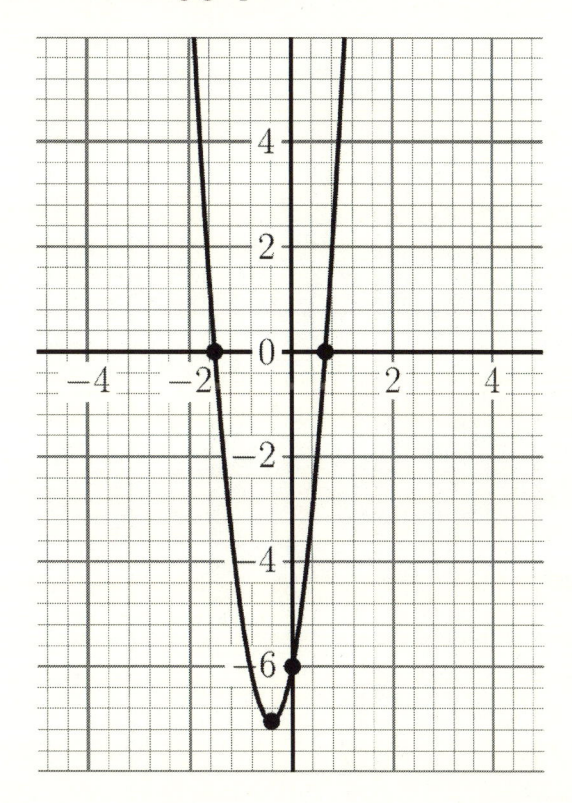

You'll probably need to use the zoom in/zoom out controls to get the scale right so you can see the 2 points where the parabola crosses the *x*-axis. Since these are the 2 points where the *y*-coordinate is 0, the *x*-coordinates at these points are the answer. Clicking on these points will confirm the answers we got before, -1.5 and 2/3 (though you do need to know that .667 is approximately 2/3).

Or let's say you need to get a quadratic into its vertex form. Earlier, we did that with $x^2 + 6x - 10$. If we go ahead and enter this into the OSC, we get the following (once we find the right zoom level).

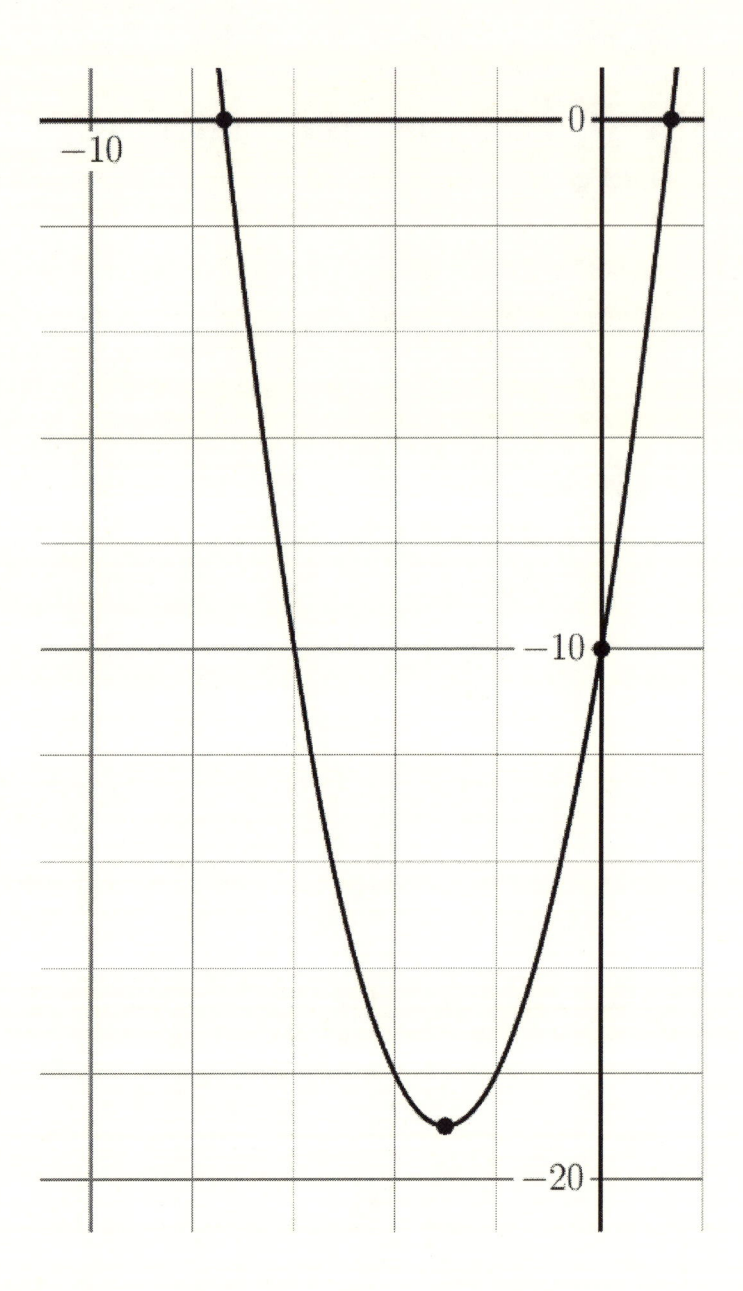

If you were to click on the vertex, you'd see that it's at $(-3, -19)$. That means the vertex form of the quadratic is $(x+3)^2 - 19$.

There will be more about the OSC later!

Day 4

Drill

*If you are confident about a topic below, you can do only the **bolded** questions. Remember, aim for **efficiency**!*

Solving Simple Quadratic Equations Set to 0: Factor the Quadratic into 2 Binomials, and Solve.

1. $x^2 + 8x + 15 = 0$
2. $x^2 - x - 110 = 0$

Solving Simple Quadratic Equations *Not* Set to 0: If Possible, *Make* It Equal to 0.

3. $x^2 + 6x = -8$
4. $a^2 - 3a = 130$

Factoring Quadratics without a Middle Term: The Difference of 2 Squares.
Factor the Following.

5. $x^2 - 144$
6. $y^2 - \frac{1}{25}$

Solving Quadratics Equations with 1 Solution: Look for a Perfect Square in the Last Term (the *Constant*).

7. $a^2 + 18a + 81 = 0$
8. $x^2 - 60x - 900 = 0$

Solving Quadratics through Grouping: Re-express the Quadratic as 4 Terms, and Then Group Them into 2 Pairs.

9. $4x^2 + 8x + 3 = 0$
10. $4b^2 - 9b - 9 = 0$

Completing the Square: Express a Quadratic in the Vertex Form $a(x - h)^2 + k$.

11. $x^2 + 4x + 7$
12. $x^2 + 50x - 5$

Solving by the Quadratic Formula: $x = \frac{-b \pm \sqrt{b^2 - 4ac}}{2a}$

13. $x^2 + 7x - 2 = 0$
14. $3x^2 + 5x - 10 = 0$

BONUS QUESTIONS

15. $x^2 + x - 20 > 0$
Solve for a range of values.

16. $12y + 9 = -3y^2$
Solve for y.

17. Simplify the following: $\frac{2}{\sqrt{6}} - \frac{1}{3}$

18. Solve: $216^{\frac{1}{3}} = -2x$

Day 4: Drill Explanations

Solving Simple Quadratic Equations Set to 0: Factor the Quadratic into 2 Binomials, and Solve.

1. $x^2 + 8x + 15 = 0 \ (x+3)(x+5) = 0 \ x = -3, -5$
2. $x^2 - x - 110 = 0 \ (x-11)(x+10) = 0 \ x = 11, -10$

Solving Simple Quadratic Equations *Not* Set to 0: If Possible, *Make* It Equal to 0.

3. $x^2 + 6x = -8 \ x^2 + 6x + 8 = 0 \ (x+2)(x+4) = 0 \ x = -2, -4$
4. $a^2 - 3a = 130 \ a^2 - 3a - 130 = 0 \ (a-13)(a+10) = 0 \ x = 13, -10$

Factoring Quadratics without a Middle Term: The Difference of 2 Squares.
Factor the Following:

5. $x^2 - 144 = (x-12)(x+12)$
6. $y^2 - \frac{1}{25} = \left(y - \frac{1}{5}\right)\left(y + \frac{1}{5}\right)$

Solving Quadratics Equations with 1 Solution: Look for a Perfect Square in the Last Term (the *Constant*).

7. $a^2 + 18a + 81 = 0 \ (a+9)^2 = 0 \ a = -9$
8. $x^2 - 60x + 900 = 0 \ (x-30)^2 = 0 \ x = 30$

Solving Quadratics through Grouping: Re-express the Quadratic as 4 Terms, and Then Group Them into 2 Pairs.

9. $4x^2 + 8x + 3 = 0 \ (4x^2 + 2x) + (6x + 3) = 0 \ 2x(2x+1) + 3(2x+1) = 0$
$(2x+3)(2x+1) = 0 \ 2x + 3 = 0 \text{ and } 2x + 1 = 0 \ x = -\frac{3}{2}, -\frac{1}{2}$

10. $4b^2 - 9b - 9 = 0 \ (4b^2 - 12b) + (3b - 9) = 0 \ 4b(b-3) + 3(b-3) = 0$
$(4b+3)(b-3) = 0 \ 4b + 3 = 0 \text{ and } b - 3 = 0 \ b = -\frac{3}{4}, 3$

Completing the Square: Express a Quadratic in the Vertex Form $a(x+b)^2 + c$

11. $x^2 + 4x - 1 = (x+2)^2 - 5$
12. $x^2 + 50x - 5 = (x+25)^2 - 630$

Solving by the Quadratic Formula: $x = \frac{-b \pm \sqrt{b^2 - 4ac}}{2a}$

13. $x^2 + 7x - 2 = 0$ $x = \frac{-7 \pm \sqrt{7^2 - 4(1)(-2)}}{2(1)}$ $x = \frac{-7 \pm \sqrt{57}}{2}$ $x = \frac{-7 + \sqrt{57}}{2}$ and $x = \frac{-7 - \sqrt{57}}{2}$

14. $3x^2 + 5x - 10 = 0$ $x = \frac{-5 \pm \sqrt{5^2 - 4(3)(-10)}}{2(3)}$ $x = \frac{-5 \pm \sqrt{145}}{6}$ $x = \frac{-5 - \sqrt{145}}{6}$ and $x = \frac{-5 + \sqrt{145}}{6}$

MORE BONUS QUESTIONS

15. $x^2 + x - 20 > 0$ Solve for a range of values. $(x - 4)(x + 5) > 0$ $x < -5$ and $x > 4$

16. $12y + 9 = -3y^2$ Solve for y. $3y^2 + 12y + 9 = 0$ $3(y^2 + 4y + 3) = 0$ $(y^2 + 4y + 3) = 0$ $(y + 1)(y + 3) = 0$ $y = -1, -3$

17. Simplify the following: $\frac{2}{\sqrt{6}} - \frac{1}{3}$ First, $\frac{2}{\sqrt{6}} \times \frac{\sqrt{6}}{\sqrt{6}} = \frac{2\sqrt{6}}{6} = \frac{\sqrt{6}}{3}$ Next, $\frac{\sqrt{6}}{3} - \frac{1}{3} = \frac{\sqrt{6} - 1}{3}$

18. Solve: $216^{\frac{1}{3}} = -2x$ First, $216^{\frac{1}{3}} = \sqrt[3]{216} = 6$ Next, $6 = -2x$ and $x = -3$

Day 5

Word Problems 1 – Rate, Average, Backsolving, Picking Numbers

After reading this material and completing the drill at the end, sign and check the box for Day 5.

Word problems are math problems that typically involve real-world elements such as prices, speeds, and distances. Many of these problems can be solved through algebra, and that's most of what we'll do today.

A simple word problem involves an unknown value such as the time it takes Robert to drive from his home to his friend's house. We can call that unknown value x or some other letter that helps us remember what that letter represents. We might call it t for time. Using a letter other than x is helpful when we have more than 1 unknown.

Since we're talking about time, let's look at one of the most useful formulas you'll need to know.

Rate x Time = Distance (RT=D for short): Once you know 2 of these values, you can easily calculate the 3rd.

This formula is one we do actually use in our lives. You might know that you have to drive a distance of 120 miles and that you will drive on average about 60 miles per hour. We probably wouldn't bother thinking about the formula for this simple situation, but you could write $60t = 120$. Dividing each side by 60 gives us $t = 2$. The drive will take about 2 hours.

This *very* simple example could become a bit more complex. For example:

> Simple: Robert plans to drive to his friend Emile's house 120 miles away.
> He'll take the highway for 100 miles and back roads for the remainder of the trip.
> His highway speed will average about 60 miles per hour,
> and his back road speed will average about 40 miles per hour.
> That includes any stops he makes.
> Which of the following will be closest to Robert's total drive time?

We'll have to use the formula twice. First, we have $60t = 100$. That means $t = \frac{100}{60} = \frac{10}{6}$ or simply $\frac{5}{3}$ of an hour. You could continue on to the 2nd leg of the trip, or you could think about what $\frac{5}{3}$ of an

Word Problems 1 – Rate, Average, Backsolving, Picking Numbers | 51

hour is. This is the same as $1\frac{2}{3}$ of an hour. Since 1/3 of an hour is 20 minutes, 2/3 is 40 minutes. And $1\frac{2}{3}$ of an hour is 1 hour and 40 minutes. (Using a calculator for this is fine as long as you work efficiently. Often, it's better to use fractions and some mental math.)

Now let's look at the 2nd leg of the trip. You *could* use t again for this other time, giving you $40t = 20$ as long as you don't get confused, or you could use a different letters such as t_1 and t_2. In any case, this 2nd time is ½ of an hour. Since that's 30 minutes, the total time is 1 hour and 70 minutes, or 2 hours and 10 minutes.

Let's make the situation more challenging.

> Challenging: Robert's friend Emile lives 230 miles away.
> They plan to meet at a restaurant along the highway that connects their towns.
> Robert will drive there at an average speed of 60 miles per hour,
> and Emile will drive there at an average speed of 55 miles per hour.
> If they both leave at noon, at what time will they meet?

Let's think about what we know and what we don't know. We know their individual rates—60 and 55. We don't know their times, and we don't know their individual distances. They won't meet in the middle since one driver is faster. However, we do know their combined distance. Once they've met, if we add their distances, we'll get 230 miles. That's their *combined* distance.

Even though we don't know their individual distances, we can *express* them. Using our mighty algebra skills and our equation RT=D, we can say that Robert's distance is $60t$. That's his rate (60) multiplied by his time (t). **This is key.** You need to get comfortable with the idea of **expressing** unknown values algebraically. What's his distance? We don't know (yet), but we can *express* it as $60t$.

What about Emile's distance? That's just $55t$. But let's be careful. We've used t in 2 different expressions. Are they really the same value? If not, calling them both t could be a problem. Did they both spend the same amount of time on the road?

Yes. The problem says they left at the same time. So when they meet, the same amount of time will have gone by. But if, for instance, Emile had left an hour after Robert, we could say that Robert's time was t and Emile's time was $t - 1$.

But okay, in this case we have the same t, so we can now say $60t + 55t = 230$. Do you see why? The sum of their individual distances is 230 miles, as we said up above. We can now easily solve this equation, and $60t + 55t = 230$ becomes $115t = 230$ and $t = 2$. In other words, they both spent 2 hours on the road. Since they left at noon, they met at 2:00 p.m.

Just for fun, we can now also easily calculate how far they each traveled—their individual distances. Robert went 60 x 2 or 120 miles, and Emile went 55 x 2 or 110 miles. Add those together, and of course, we get 230 miles.

Since RT = D, we can say that R = D/T and T = D/R. It's all the same info. So if you know that someone traveled 500 feet, you can say that their Rate was 500/T or their Time was 500/R. Get comfortable with this 3-part formula and also with the next one.

Average = Sum/Number of Terms: Once You Know 2 of These Values, You Can Easily Calculate the 3rd (sound familiar?).

This could be the most useful formula to know on the test. And you probably know it pretty well, though you might not know all the ways it can be used. Let's first look at a very simple application of this formula. (On the test, *average* is often referred to as *arithmetic mean*. They're the same thing.)

> Simple: The arithmetic mean of the test results of a class of 20 students was 85. One student got a 95.
> What was the sum of the other 19 scores?

Using the formula, we can say that $85 = \frac{Sum}{20}$. Note that the average was **given**. It is not an unknown. The SAT authors often do this. What's unknown is the **sum**, which we can easily calculate by multiplying 85 by 20, giving us 1,700. If you added up all the scores, that's what you'd get. Since 1 student got a 95, the sum of the other scores is 1,700 – 95, which is 1,605.

> Challenging: Set A consists of 5 numbers, 7, 10, 20, 25, and *x*.
> The arithmetic mean of the 5 numbers in Set A is half the arithmetic mean of the 4 numbers in Set B.
> If the sum of the numbers in Set B is 112, what is the value of *x*?

We know the most about Set B. It has 4 numbers and a sum of 112. To get the mean, we just divide 112 by 4, giving us 28. A's mean is half of this, which would be 14. Now that we know that, we can multiply 14 by 5 (average x number of terms) to get the sum, which is 70. The sum of 7, 10, 20, and 25 is 62, so *x* must equal 8 (62 +8 = 70). We just keep applying the same formula (and simple arithmetic) until we get the answer.

Word Problems with 2 Unknowns: Set Up 2 Equations

We've looked at how to handle *simultaneous equations*—for example, 2 equations with 2 unknowns. Often, you'll create 2 equations in a word problem, assigning each unknown a different letter.

> Simple: Sal spends $8.40 at the grocery store buying bunches of bananas priced at $1.10 per bunch and cans of soup that cost 85 cents each.
> There is no tax on the purchase.
> If the total number of bunches of bananas and cans of soup is 9, how many cans of soup did Sal buy?

The trick here is to use letters to represent unknowns and to be **absolutely clear** about what those letters represent. For example, you can't say, "I'm going to use *b* to represent bananas. But what about those bananas—their weight, cost, the number purchased? Since we don't know how many bunches were bought, we could use *b* to represent this. You might even write, "*b* = number of bunches." Likewise, you could use *c* to represent the number of cans of soup purchased. Now we can write our **2** equations. (Remember, for 2 unknowns, you'll need 2 equations.) We know that Sal bought 9 items total, so you could write $b + c = 9$. Our other equation involves the cost. That would be $1.10b + .85c = 8.40$. You need to be absolutely secure in setting up these equations. If you're not, take another look at this material.

We need to solve for *c,* the number of cans bought. So with $b + c = 9$ and $1.10b + .85c = 8.40$, we'll use either substitution or elimination to solve it. Since we want to solve for *c*, we want one equation that only has *c* in it. We'll take the 1st equation and get *b* in terms of *c*: $b = 9 - c$. Now we'll substitute this information into the 2nd equation: $1.10(9 - c) + .85c = 8.40$. Great! That's 1 equation with 1 unknown. Once we distribute that 1.10, we get $9.90 - 1.10c + .85c = 8.40$. When we combine like terms, we get $9.90 - .25c = 8.40$. If we add $.25c$ to each side and subtract 8.40 from each side, we get $1.50 = .25c$, and then $c = 6$.

> Challenging: At a warehouse, 5,900 pounds of furniture boxes are being moved. Some of the boxes weigh 320 pounds each, and the rest weigh 540 pounds each.
> If there are twice as many of the lighter boxes, how many of the lighter boxes are being moved?

Again, there are 2 equations for 2 unknowns. You might be able to combine them mentally and write it out as 1 equation, and that's fine if you're accurate. But we'll break it down here. We'll let *l* represent the number of light boxes and *h* the number of heavy boxes. Since there are twice as many of the lighter boxes, we can say that $l = 2h$. Knowing that $320l + 540h = 5900$, we can use simple substitution to say that $320(2h) + 540h = 5900$. This simplifies to $640h + 540h = 5900$, and then $1180h = 5900$, and finally, $h = 5$.

Be careful! We solved for *h*, but the question asked for the number of lighter boxes. That's easy. We know there were twice as many of the lighter ones, so $l = 10$. Do you think that 5 will show up among the wrong choices? We do too.

However, let's think about this. The SAT folks include *likely* wrong choices, not just random numbers. So it's likely that among the choices *for this* question, there will be a wrong choice that is half the correct choice. Let's say that the choices were 2, 5, 10, and 12. If you had to take a quick **intelligent** guess, 10 would be a good one. Of course, they could have made it tougher by having the choices as 2, 5, 10, and 20. Still, in this case, you could feel reasonably confident narrowing the choices down to 10 and 20. Yes, it's better to work the thing out correctly, but it's also an excellent idea to scan the choices, especially when you're short on time. (More on this later.)

If you were tempted to try out the choices rather than set up the equations, that can be an effective approach. We'll discuss that in depth in just a bit. But you **should** be able to set up the equations efficiently and solve it. Have as many tools in your tool kit as you can.

Word Problems with Quadratics: Set to Aero, If Possible, Know the Vertex Formula, and Consider Graphing

Simple: A researcher knows that as she gradually adds more of Substance A to water, it will decrease the mixture's temperature until it begins to have the opposite effect, increasing the mixture's temperature. She determines that this relationship can be defined as $\Delta t = a^2 - 9a + 20$ where Δt is the change in temperature and a is the amount of Substance A added to the water in grams. For what positive value or values would the number of grams of the substance added to the water neither increase nor decrease the mixture's temperature?

In other words, at what point will Δt be 0, meaning **no** change to the temperature? That does, indeed, allow us to set the quadratic to 0: $0 = a^2 - 9a + 20$. From here, we do what we often do with quadratics. We factor. This gives us $(a-4)(a-5) = 0$, which means that $a = 4$ and $a = 5$. So the answer would be "both 4 ounces and 5 ounces." An answer choice that said "only 4 ounces" would be incorrect.

You might have been tempted to use your calculator (remember, you'll have an on-screen calculator [OSC] during the test) to graph the quadratic and look for the 0's, the x coordinates of the points where the parabola crosses the x-axis, since quadratics give us parabolas when graphed. That's a fine way to do it **if** you can do so efficiently. But this problem can be solved easily with factoring. We'll talk about using the calculator more on another day, but remember that when using the calculator, use x or y, not (as in this example) a or some other letter.

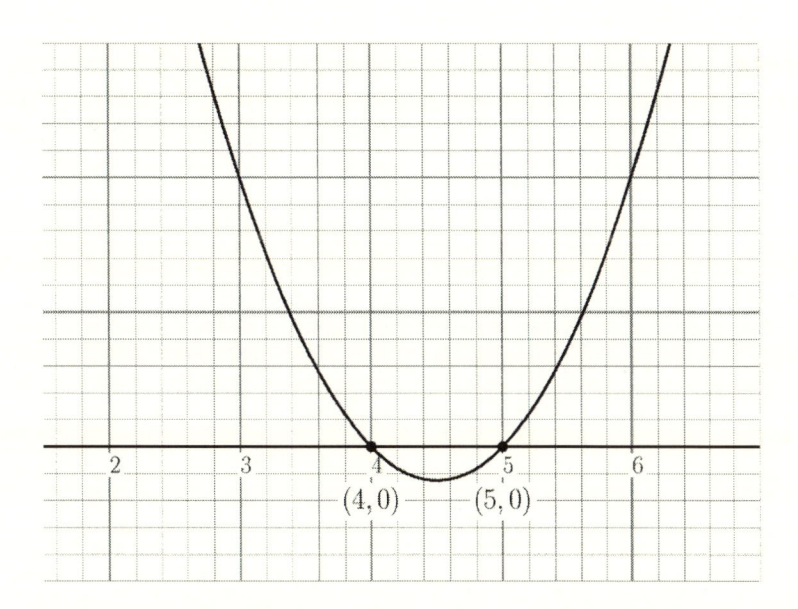

Challenging: The designers of a new car engine have determined that the efficiency (e) of the engine increases as its speed in miles per hour (s) increases until a point where an increase in speed decreases its efficiency. The designers have determined that $e = -\frac{5}{2}s^2 + 400s - 30$ where $s > 0$, describes this relationship. At what speed is the engine most efficient?

Word Problems 1 – Rate, Average, Backsolving, Picking Numbers | 55

Here are a few things to notice right off the bat: (1) there is no reason to set the quadratic to 0, and (2) the coefficient of the first term is negative, which, as you might know (or remember from the quadratic lesson on Day 3) means that this quadratic, when graphed, gives us a downward opening parabola. That makes sense for this situation. As our x value (speed) increases along the axis from left to right, the y value (efficiency) increases until it reaches its maximum and then heads down. What we want is the speed (x) that gives us that maximum y. This highest point of the parabola is the vertex.

Here, the on-screen calculator is helpful, but remember to replace s with x, and replace e with y. You would then look for the x value of the vertex—the highest point—though you'll have to zoom in or out to get the scale to be readable.

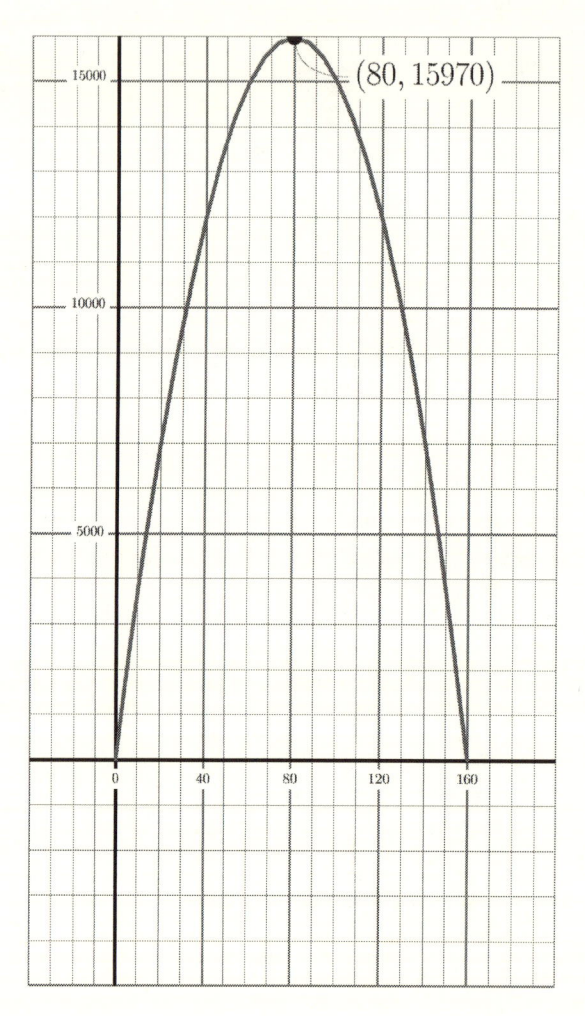

However, there's also a simple formula that gives you the x value of the vertex: $x = -\frac{b}{2a}$. So, for quadratic $-\frac{5}{2}s^2 + 400s - 30$, $a = -\frac{5}{2}$ and $b = 400$. Using these values, we have $x = -\frac{400}{2\left(-\frac{5}{2}\right)} = \frac{400}{5} = 80 \ mph$. That's the speed at which the engine is running at maximum efficiency.

But here's something to keep in mind. You might remember—or maybe you already knew—that quadratics can be expressed in the form $a(x - h)^2 + k$, which is known as the **vertex** form because it just serves us the vertex on a platter. The coordinates of the vertex are (h, k). For example, on Day 4, we looked at a quadratic in its vertex form: $(x + 3)^2 - 19$. That tells us at a glance that the vertex of this parabola, were we to graph it, is $(-3, -19)$. Note that the x coordinate of the vertex is the opposite of the number **added** to x. If we had the quadratic $(x - 9)^2 + 4$, the vertex would be $(9, 4)$. So if you have a quadratic in the vertex form and you need the formula, the coordinates are **right there** in the expression.

Backsolving: When the Choices Contain Known Values

So far, we haven't looked at questions with the 4 multiple choices you'll see on most SAT questions. Examining these choices can lead us to the correct one, **sometimes** with greater efficiency than if we had worked the problem out in the "traditional" way. But be careful. Frequently, students try this approach as often as they can to avoid the algebra/arithmetic involved. You need those basic skills! But sure, at times **backsolving** the alternative approach that involves trying out the choices is a good approach. Let's look at that.

You can backsolve when there are **only known values** in the choices. If you have unknown values such as x, we have another alternative strategy that we'll look at next.

Simple: There are 32 houses on Piedmont Street,
some of which have solar panels.
A local company has rated the energy efficiency of each of these houses
on a scale of 1 to 10. The houses with the panels received
a rating of 8, and the others received a rating of 3.
The combined efficiency rating of these houses is 206.
How many of the houses have solar panels?

A) 10
B) 12
C) 20
D) 22

We'll look at the traditional algebraic approach in a moment, but let's try to backsolve this from the choices. Usually, it's best to start with 1 of the 2 center choices, B and C. If you try B and it's too high, then the answer is A. If you try C and it's too low, the answer is D. However, it's not always obvious whether a choice is too low or too high. In our case, let's try choice B, which is 12. Here is what the question is now.

There are 32 houses on Piedmont Street; **12** of them have solar panels, and 20 do not. The 12 houses with panels were each rated 8, giving them a total rating of 96. The other 20 each got a rating of 3, giving them a total rating of 60. The combined efficiency rating of these houses is 96 + 60, which is 156, and that means choice B is **not** correct because the actual combined rating is 206.

Since we want a higher combined rating, it makes sense to try choice C or D because we want more of the high-efficiency houses. Let's try C.

There are **20** houses that have solar panels, and 12 do not. The 20 houses with panels were each rated 8, giving them a total rating of 160. The other 12 each got a rating of 3, giving them a total rating of

Word Problems 1 – Rate, Average, Backsolving, Picking Numbers | 57

36. The combined efficiency rating of these houses is 160 + 36, which is 196, and that means that choice C is **not** correct because the actual combined rating is 206.

It also means that we went in the right direction, so choice D must be correct. If you're confident that you've backsolved correctly, go ahead and choose D, but let's confirm it here.

There are **22** houses that have solar panels, and 10 do not. The 22 houses with panels were each rated 8, giving them a total rating of 176. The other 10 each got a rating of 3, giving them a total of 30. The combined efficiency rating of these houses is 176 + 30, which is 206, and that means that **choice D is correct.**

Here's the algebraic approach, using 2 equations for our 2 unknowns. We'll let s represent the number of houses with solar panels and n represent the ones that do not.

$s + n = 32$ and $8s + 3n = 206$. Using substitution, we can say that $n = 32 - s$ and $8s + 3(32 - s) = 206$.

Finally, after distribution of the 3 and collecting like terms: $5s = 110$, and $s = 22$.

Some of you might prefer 1 approach over the other for this question, but you **should** be able to use either one efficiently. There are definitely questions where the traditional approach would be more efficient, and of course there are the math questions that give you **no choices**. But backsolving can come in handy, especially for the later, more time-consuming questions.

> Challenging: A school district has 260 4th-grade students.
> Next year, there will be one fewer 4th-grade class,
> and the average class size is expected to be 2 more than it is this year,
> making the total number of 4th-grade students 264.
> What is the current average class size?

Again, this **can** be done algebraically. In fact, let's at least set it up that way. We'll use m to represent the current average (mean) of the classes—that's what we're looking for. We'll use c to represent the current number of classes. We can then use our average formula to create these 2 equations: $mc = 260$ and $(m + 2)(c - 1) = 264$. We can now use substitution, but it's going to get a little intricate. If we use substitution, we can say that $m = \frac{260}{c}$, but plugging this into the other equation will lead to a quadratic, and though it's doable, there are enough steps that errors can occur. So let's backsolve. Let's say our choices are the following:

A) 12
B) 13
C) 20
D) 22

Let's start with B this time. That means the current average class size is (m) = 13. It also means the average class size next year is expected to be 15 (2 more). It's actually helpful at this point to have set up the algebraic equations as shown above. Since the average times the number of classes is 260 ($mc = 260$), we know that if m is 13, then $c = 20$. If we use these values in the 2nd equation, it doesn't work.

58 |Day 5

$(13 + 2)(20 - 1)$ equals 285, **not** 264. This is an example of when you might not be sure what to try next. Sometimes it just makes sense to try a simple value. Let's go with choice C: 20.

If m is 20, then $mc = 260$ tells us that c is 13. And $(20 + 2)(13 - 1)$ does indeed equal 264. **C is correct.**

What if you had tried B and then tried A? That wouldn't be too bad because it would mean that c, the number of classes, would not be an integer. That can't be. The same thing would occur when you tried choice D. This rules out both of those choices.

Picking Numbers: When the Choices Contain Unknown Values

When the choices contain unknown values such as x, consider picking your own numbers for these values. We did this when discussing percentage problems where 100 is often a good value to choose for an unknown such as the cost of an item. But this approach can be used elsewhere, but often not in a classic real-world word problem such as the following:

> Simple: If x and y are consecutive odd integers,
> with $y > x$, and $x + 2y = z$,
> which of the following is equivalent to z?
>
> A) $3y - 2$
> B) $3y + 2$
> C) $6y - 2$
> D) $6y + 4$

We can't choose any numbers we want; we need 2 consecutive **odd** integers, with y the larger one. It's usually a good idea to avoid choosing the number 1 (it might lead to 2 choices that seem correct). So let's go with $x = 3$ and $y = 5$. In that case, $x + 2y$ is 13. In other words, z is 13. The question wanted a choice equivalent to z, so all we have to do is plug 5 in for y until we get 13. That won't take long. Choice A gives us 13. **A is correct.**

Let's introduce a little geometry (the subject of tomorrow's lesson) for this next one.

> Challenging: If the volume of Cube B is four times the volume of Cube A,
> the length of one edge of Cube B is
> how many times the length of one edge of Cube A?
>
> A) $\sqrt[3]{4}$
> B) 2
> C) 3
> D) $3\sqrt{2}$

Here's an example of picking numbers when the choices are actually **known** values, though they represent not a value such as 2 but rather 2 *times* another value. As you may know, the volume of a

Word Problems 1 – Rate, Average, Backsolving, Picking Numbers | 59

cube is equal to 1 of its edges **cubed** ($V = e^3$). Let's pick 2 for the smaller volume and therefore 8 for the larger volume. (Technically, those are *cubic units, like 2 cubic inches and 8 cubic inches*.)

Since the smaller cube has a volume of 2, its edge is $\sqrt[3]{2}$. The larger cube has a volume of 8, so its edge length is simply 2. If we divide 8 by $\sqrt[3]{2}$, we'll know how many times longer the edge of the bigger cube is. As a fraction, that looks like $\frac{8}{\sqrt[3]{2}}$. This is the same as saying $\frac{2^3}{2^{\frac{1}{3}}}$. Now that we have the same base on the top and bottom, we can **keep that base and subtract the exponents**. That gives us $2^{\frac{2}{3}}$. The answer might be written this way, but since we don't see it in the choices, we have to go a little further. Raising 2 to the 2/3 power is the same as squaring 2, giving us 4, and then taking the cubed root of that, giving us **choice A**.

If that seemed really time-consuming, we could have arrived at the right choice earlier in the process. If you use the calculator to divide 8 by $\sqrt[3]{2}$, you'll see a decimal beginning with 1.587. That is correct but not in the form of 1 of the choices. **However,** it's clearly not choices B and C, and it's also not D, which is greater than 3. If you wanted to be absolutely sure, you could also use the calculator to determine the value of choice A, which you could do by entering $4^{\frac{1}{3}}$. Sure enough, you'll see the same decimal.

Day 5

Drill

If you are confident about a topic below, you can do only the **bolded** questions. Use all the choices to help you arrive at the correct one. If you think that graphing, backsolving, or picking numbers is the best approach, that's fine. But consider algebraic solutions. To really make use of this drill, try to solve the answers several ways. You should know them all!

Rate x Time = Distance: Once You Know 2 of These Values, You Can Easily Calculate the 3rd

1. A roller coaster ride lasts 2½ minutes, during which each car travels 3,100 feet. What is the average speed of each car in feet per second?

   A) 20
   B) $20\frac{2}{3}$
   C) 21
   D) $21\frac{1}{3}$

2. Every day, a train travels 270 miles from Wardsdale to Everett City. Once construction along the track is completed, the train will be able to travel this same distance 9 miles per hour faster, which means it will save 1 hour of travel time. What is the train's current speed traveling between Wardsdale and Everett City?

   A) 44
   B) 45
   C) 54
   D) 55

Average = Sum/Number of Terms. Once you know 2 of these values, you can easily calculate the 3rd. (Sound familiar?)

Drill | 61

3. At a 3-day festival, the average daily attendance was 310. The next year, the festival was extended to 4 days, during which the average daily attendance was 540. What was the positive difference between the total attendance for the 2 festivals?

4. Alain, Bonnie, and Malik are the only 3 salespeople who work at a car dealership that sells both new and used cars. During 1 month, sales of cars totaled $1,960,000, with an average selling price of $28,000. Alain sold 22 cars, and Bonnie sold 25. If Malik sold both new and used cars that month, what is the maximum number of used cars that Malik could have sold?

A) 1
B) 22
C) 23
D) 45

Word Problems with 2 Unknowns: Set Up 2 Equations

5. A bookstore purchased 37 books for a total of $445.00. Some of the books cost $10.00 each, and the others cost $15.00 each. How many of the more expensive books were purchased?

6. An architect is designing 2 office buildings that both have the same number of stories (floors). The height of each story in 1 of the buildings is 12 feet, and the height of each story in the other building is 13 feet. There is no height to the buildings other than the stories. If the combined height of the buildings is 450 feet, what is the height of the shorter building in feet?

A) 25
B) 192
C) 216
D) 234

Word Problems with Quadratics: Set to 0, If Possible, Know the Vertex Formula, and Consider Graphing

7. A business has determined that the profit (p) made selling a toy for s dollars can be estimated using the equation $p = s^2 - 18s + 80$.

A negative *p* value means the company will lose money selling the toy. Which of the following selling prices would result in a loss? (Ignore the dollar sign in your answer.)

Note: This is a question where you need to enter your answer, not choose from 4 choices.

8. In a research lab, a mouse completed a maze that was 30 feet in length. The next day, its speed completing the maze was 0.5 feet per second faster than on the previous day, allowing it to complete the maze in 2 seconds less time than on the previous day. What was the average speed on the 1st day?

BONUS QUESTIONS

9. The employees at a real estate firm were rated on their annual performance using a scale of 1 to 20. All scores were integers. The average rating 1 year was 15, and the sum of all the scores was 225. If 3 of the employees got the highest possible score, how many did not receive the highest possible score?

A) 12
B) 15
C) 17
D) 19

10. A car service charges a certain amount for every mile traveled as well as a fixed fee for any ride. The equation $d = 3.50m + 3$ represents the total amount, *d*, in dollars for a ride of *m* miles. What does 3 represent in the equation?

A) The cost of a ride of 1 mile in dollars
B) The cost of a ride of any number of miles in dollars
C) The amount of the fixed fee in dollars
D) The distance the car service will travel for $3.50 in miles

11. Aya bought a pair of headphones for $69. She later found out that this price was 15% more than it was at another location. What was the price at the other location?

A) $45.00
B) $54.00
C) $60.00
D) $79.00

Drill | 63

12. The ratio of the number of cats to dogs seen at a clinic 1 year was 17:20, and the total number of cats and dogs seen that year was 740. What was the number of cats seen at the clinic that year?

A) 60
B) 340
C) 370
D) 400

Day 5

Drill Explanations

Rate x Time = Distance: Once You Know 2 of These Values, You Can Easily Calculate the 3rd

1. A roller coaster ride lasts 2½ minutes, during which each car travels 3,100 feet. What is the average speed of each car in feet per second?

   A) 20
   B) $20\frac{2}{3}$
   C) 21
   D) $21\frac{1}{3}$

Time (t) = 150 seconds. We're using seconds because the question wants the answer in feet per **second**.

Distance (d) = 3100 feet. To get rate (speed), we just calculate distance divided by time, which is $\frac{3100}{150} = 20\frac{2}{3}$ feet per second. If you used your calculator, you need to be able to recognize that .66666 repeating or .666667 is 2/3.

2. Every day, a train travels 270 miles from Wardsdale to Everett City. Once construction along the track is completed, the train will be able to travel this same distance 9 miles per hour faster, which means it will save 1 hour of travel time. What is the train's current speed traveling between Wardsdale and Everett City?

   A) 44
   B) 45
   C) 54
   D) 55

This can be done algebraically, but backsolving works well. Sometimes, you do a little of both. Let's set up the equations first. We can say that the current time (t_c) is equal to the distance (270) divided by the current rate (r). So that's $t_c = \frac{270}{r}$. The future time (t_f) will be equal to the distance (270) divide by

Drill Explanations | 65

the future rate, which is going to be $r + 9$, so that's $t_f = \frac{270}{r+9}$. And then, since the future time will be 1 hour less, we can say $\frac{270}{r+9} = \frac{270}{r} - 1$. This can be solved, but it's a little tricky. Errors can find their way into the calculating, so let's try some backsolving since our choices are **known** values. And let's be strategic. We know there's a 9 mph difference between the 2 rates, and we want the slower one. Choice B is 9 less than choice C, so that makes B a likely correct choice, and C is a likely **incorrect** choice. Let's try putting 45 into the equation we've created.

$\frac{270}{45+9} = \frac{270}{45} - 1$ becomes $\frac{270}{54} = \frac{270}{45} - 1$, and then, with a little help from the OSC, this works: $5 = 5$. We're done here. B is correct.

Average = Sum/Number of Terms: Once You Know 2 of These Values, You Can Easily Calculate the 3rd

3. At a 3-day festival, the average daily attendance was 310.
The next year, the festival was extended to 4 days,
during which the average daily attendance was 540.
What was the positive difference between the
total attendance for the 2 festivals?

A) 230
B) 850
C) 1080
D) 1230

Since the average was 310 during the 3 days, that gives us a sum of 930; that's the total attendance during the 3 days. (Remember: **average times number of terms equals sum**.) Since the average was 540 during the 4 days, that sum was 2,160. The difference between 2,160 and 930 is 1,230.

4. Alain, Bonnie, and Malik are the only 3 salespeople
who work at a car dealership that sells both new and used cars.
During 1 month, sales of cars totaled $1,960,000,
with an average selling price of $28,000.
Alain sold 22 cars, and Bonnie sold 25.
If Malik sold both new and used cars that month,
what is the maximum number of used cars that Malik could have sold?

A) 1
B) 22
C) 23
D) 45

We can get the number of cars sold by dividing—with our calculator—the total (sum) by the average. That's $\frac{1,960,000}{28,000} = 70$. Since Alain and Bonnie sold a total of 47 cars, Malik sold the remaining 23.

We're told that he sold both new and used cars, so the maximum number of used cars he could have sold is 22.

Word Problems with 2 Unknowns: Set Up 2 Equations

5. A bookstore purchased 37 books for a total of $445.00. Some of the books cost $10.00 each, and the others cost $15.00 each. How many of the more expensive books were purchased?

A) 12
B) 15
C) 22
D) 25

Let's call the number of less expensive books l and the number of the more expensive books m. Our 2 equations are $l + m = 37$ and $10l + 15m = 445$. Since we want the number of the more expensive books, let's get l in terms of m: $l = 37 - m$. Now, using substitution, we can say $10(37 - m) + 15m = 445$. After distributing the 10 and collecting like terms, we get $m = 15$. Done! If you backsolved efficiently, that's fine too.

6. An architect is designing 2 office buildings that both have the same number of stories (floors). The height of each story in 1 of the buildings is 12 feet, and the height of each story in the other building is 13 feet. There is no height to the buildings other than the stories. If the combined height of the buildings is 450 feet, what is the height of the shorter building in feet?

A) 25
B) 192
C) 216
D) 234

If the number of stories in each building is n, then we can say $12n + 13n = 450$. Note that we can only do this because we are told that the buildings have the **same** number of stories. This equation simplifies to $25n = 450$ and $n = 18$. Since each story is 12 feet in the shorter building, its combined height is 18 x 12, which is 216 feet. Backsolving could be used here as well.

Word Problems with Quadratics: Set to 0, If Possible, Know the Vertex Formula, and Consider Graphing

7. A business has determined that the profit (p) made selling a toy for s dollars can be estimated using the equation $p = s^2 - 18s + 80$.

Drill Explanations | 67

A negative *p* value means that the company will lose money selling the toy. Which of the following selling prices would result in a loss? (Ignore the dollar sign in your answer.)

Note: This is a question where you need to enter your answer, not choose from among 4 choices.

A positive *p* value means there's a profit, and a negative *p* means there's a loss. Let's find a value for *s*, the selling price, that would make *p* a negative number; that is, less than 0. Here's the inequality that would give us a negative *p* value.

$s^2 - 18s + 80 < 0$. To solve, let's factor $(s - 10)(s - 8) < 0$. For this product to be negative, then either $(s - 10)$ or $(s - 8)$ must be negative, and the other one must be positive. If *s* is greater than 10, they'll both be positive. And if *s* is less than 8, they'll both be negative. Therefore, we want any number between 8 and 10. Any number will do, such as 8.10, 8.50, or 8.70.

You could also use the on-screen calculator to graph and solve this. Here's the graph of the inequality, using *x* for *s*. You'll see that the range of acceptable *x*-values is between 8 and 10.

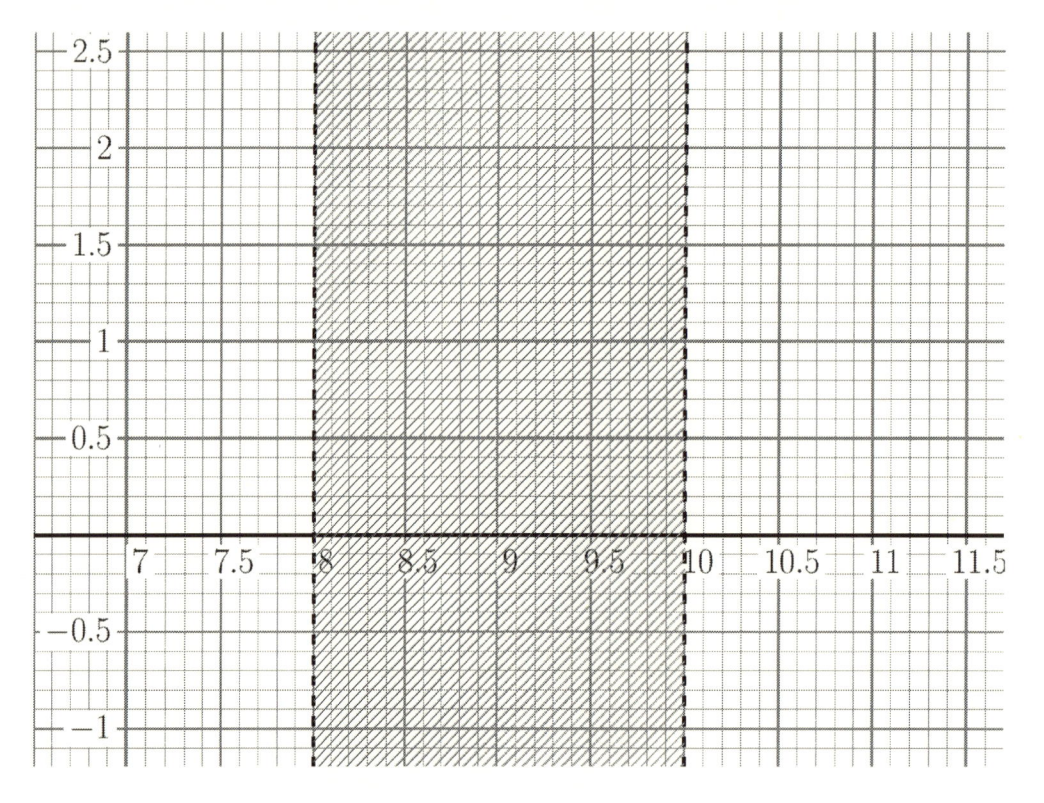

You might have chosen to graph an **equation** (not inequality) instead: $x^2 - 18x + 80 = 0$. This gives us an upward opening parabola. The scale might look odd on the calculator, but you'll see that the 0's, where the parabola crosses the *x*-axis, are at 8 and 10. A number between 8 and 10 would give us a negative *d*.

BONUS QUESTIONS

9. The employees at a real estate firm were rated on their annual performance using a scale of 1 to 20. All scores were integers. The average rating one year was 15, and the sum of all the scores was 225. If 3 of the employees got the highest possible score, how many did not receive the highest possible score?

A) 12
B) 15
C) 17
D) 19

What we need here is the total number of employees. Since we know that **the number of terms equals sum divided by average** (from the formula Average = Sum/Number of Terms), we can get the number of terms (employees) by dividing 225 by 15, giving us 15; there are 15 employees. Since 3 got the highest score (20), the other 12 did not.

10. A car service charges a certain amount for every mile traveled as well as a fixed fee for any ride. The equation $d = 3.50m + 3$ represents the total amount, d, in dollars for a ride of m miles. What does 3 represent in the equation?

A) The cost of a ride of 1 mile in dollars
B) The cost of a ride of any number of miles in dollars
C) The amount of the fixed fee in dollars
D) The distance that the car service will travel for $3.50 in miles

This linear equation is in the usual form of $y = mx + b$, which we'll discuss in more depth on a later day. Since m is the number of miles, and the service charges a certain amount **per mile**, 3.50 must refer to a rate of $3.50 for every mile. The 3 is a constant; it's not multiplied by anything. That means the 3 must be the amount of the fixed fee added to any ride.

11. Aya bought a pair of headphones for $69.00. She later found out that this price was 15% more than it was at another location. What was the price at the other location?

A) $45.00
B) $54.00
C) $60.00
D) $79.00

We can set up an equation, but be careful. Students often don't get this right. We are **not** taking 15% of 69. The 15% is added to the **other** lesser price, which is unknown. The most efficient equation would

Drill Explanations | 69

look like this, with p representing the lower price: $1.15p = 69$. This says that p **plus** 15% of p will give us 69.

Now we only need to divide each side of the equation by 1.15 to get $p = 60$. Backsolving would be fine here, too, but again, know how to set up the equations efficiently!

12. The ratio of the number of cats to dogs seen at a clinic one year was 17:20, and the total number of cats and dogs seen that year was 740. What was the number of cats seen at the clinic that year?

    A) 60
    B) 340
    C) 370
    D) 400

Remember that ratios can be expressed as fractions. If the number of cats and dogs is represented by c and d, we can say $\frac{c}{d} = \frac{17}{20}$. Our other equation is $c + d = 740$. Substitution will work here, but let's first look at the situation and the choices for clues. Since the ratio is 17 to 20, there won't be a huge difference between the number of cats and dogs, and the number of cats will be the smaller number. There are 740 of these animals all together, so we want a number less than half, but not **that** much less than half. Exactly half of 740 is 370 (trap choice C), and 60 is much too low, so 340 should look real good. You could always test it out if you have time. If the number of cats is 340, then there are 400 dogs. What is the ratio of 340 to 400? It's $\frac{340}{400}$, which simplifies to $\frac{17}{20}$. **Choice B is correct.**

Day 6

Planar Geometry – Polygons, Circles, 3-Dimensions, Right Triangles

---

After reading this material and completing the drill at the end, sign and check the box for Day 6.

There are 2 types of **geometry** that you'll need to master for this test: **planar** and **coordinate**. Today, we'll look at planar.

Planar Geometry: Know the Area and Volume Formulas

Area of a parallelogram = Base x Height, or bh
Area of a triangle = ½ bh
Area of a trapezoid = (Average of b) x h

Let's look at **parallelograms**—4-sided shapes (quadrilaterals) that have opposite sides that are parallel and equal. Most students think of a "slanty" parallelogram such as this:

This is indeed a parallelogram, but so are rectangles, including squares. **However**, the base and height of a parallelogram are always at right angles to each other, so we cannot multiply sides to get the area **unless** it is a rectangle. Otherwise, our height (also called *altitude*) would look like this:

With this parallelogram, the product of two adjacent sides is **not** the area, although it **is** likely to show up as a trap incorrect choice.

Now we'll look at **triangles**, the most important shape for this test (and perhaps for math in general). We'll especially look at **right triangles**. Once again, we have to make sure our base and height are perpendicular to each other (at right angles). With a right triangle, you can think of the legs as the base and height, but otherwise you'll need an altitude such as this one.

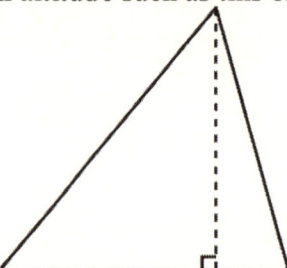

And now let's look at **trapezoids**. Yes, you can break them up into right triangles and a rectangle to get the area, but the actual formula is pretty simple and even intuitive. This time, we have **2** bases, so instead of choosing 1 of them to multiply by the height, we take the average of the 2 bases.

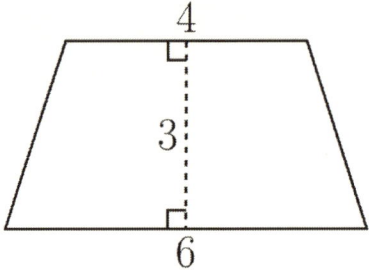

Here, the bases are 4 and 6, and the average of those 2 numbers is 5. Since the height is 3, we multiply 5 by 3 to get an area of 15 (square units). The formula is usually written as $\frac{b_1+b_2}{2} \times h$, but that's the same thing, isn't it?

Let's move on to **circles**. There are 4 things to know about circles: radius, diameter, circumference, and area. If you know one of these, you can calculate the others.

The **radius** is measured from the center to any point along the boundary.
The **diameter** is twice the radius.
The **circumference** is equal to the diameter times π. This can be written as πd or $2\pi r$. We think the first way is clearer and doesn't lead to confusion with area.
The **area** of a circle is πr^2.

The number π is an irrational number usually approximated as 3.14, although on the test, we **usually leave it as π**.

72 |Day 6

$$A = 25\pi$$
$$C = 10\pi$$
$$D = 10$$
$$r = 5$$

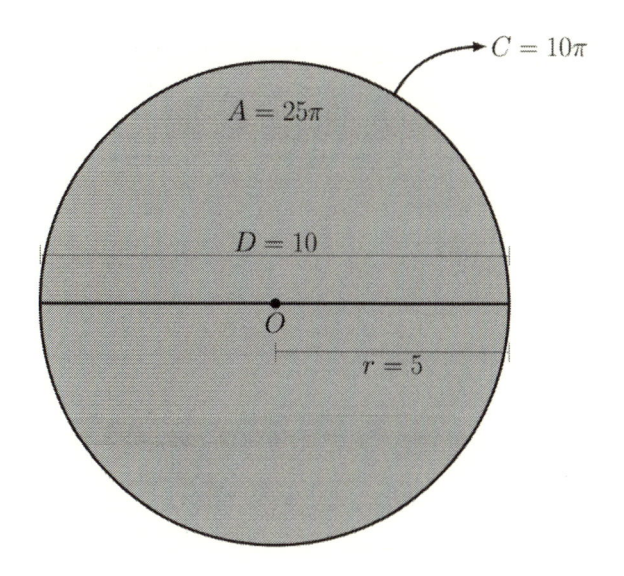

Three-dimensional shapes can show up on the test, mostly the following two.

Rectangular solids are things such as a book or a box. Besides having surface area, it has volume, like all 3-D objects. To get the volume, you just multiply the 3 dimensions—length, width, and height. In other words, $V = lwh$.

Right circular cylinders have a circular base and straight sides like a length of pipe. To get the volume, multiply the area of the base times the height. In other words, $V = \pi r^2 \times h$.

Note: It is true that you have access to a "reference" page during the test. By clicking that button, you'll see a number of formulas, including the ones we've been discussing. That's nice, but it's better to memorize them.

Angles: Memorize $180(n-2)$

You might know that a triangle and a straight line both have 180 degrees. Also, you might know that a quadrilateral (4-sided shape) and a circle both have 360 degrees. The formula above will give you the sum of the angles in an *n*-sided shape. For example, a 12-sided shape has angles that sum to 1800 degrees because 180 x 10 = 1800.

Now let's take a closer look at the all-important right triangles.

Right Triangles: Know the Specials

The formula $a^2 + b^2 = c^2$ is of great importance to geometry. This equation, the Pythagorean theorem, tells us that if you square each of the 2 legs of a right triangle and then add those squared values, you'll get the square of the hypotenuse, the longest side, the one across from the 90-degree angle. However, you can often avoid using this formula if you memorize a few "special" right triangles.

The 45/45. This is a right isosceles triangle. In other words, the two legs have the same length, and the angles are 45, 45, and 90 degrees. If one leg is 5 inches, the other leg is 5 inches, and the hypotenuse is $5\sqrt{2}$ inches. We can say that the ratio of the 3 sides is $1 : 1 : \sqrt{2}$. It can also be useful to think of it as the $x, x, x\sqrt{2}$ triangle.

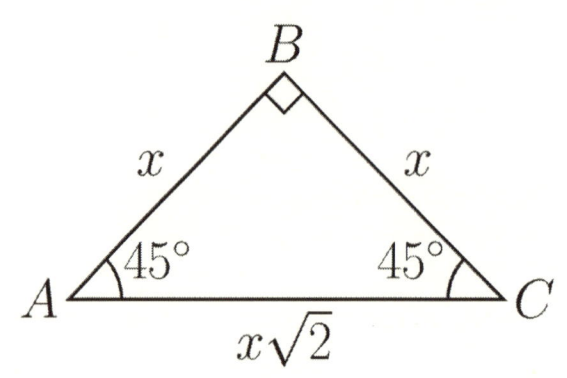

The 30/60. The angles are 30, 60, and 90 degrees. In this one, the hypotenuse is **twice the length of the short leg**. The longer leg is $\sqrt{3}$ times the short leg. We can say that the ratio of the 3 sides is $1 : \sqrt{3} : 2$. It can also be useful to think of it as the $x, x\sqrt{3}, 2x$ triangle.

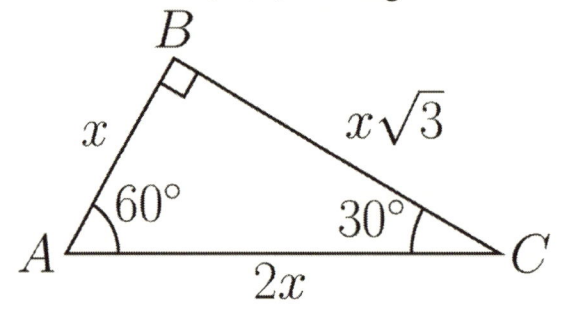

It's also wise to know some "Pythagorean triples." These are right triangles whose sides are integer values. The best known one is the **3 : 4 : 5.** If you see a right triangle with legs 3 and 4, you know that the hypotenuse is 5. And if you see a right triangle with legs 6 and 8, you know that the hypotenuse is 10 because these values are twice 3, 4, and 5. If you see a right triangle with a leg of 60 and a hypotenuse of 100, you would know that the other leg is 80. That's also a 3, 4, 5 right triangle.

Another triple has sides in the ratio **5 : 12 : 13**. So if you see a right triangle with legs 10 and 24, you would know that the hypotenuse is 26. Of course, you could use the Pythagorean theorem, but if you can save some time, why not?

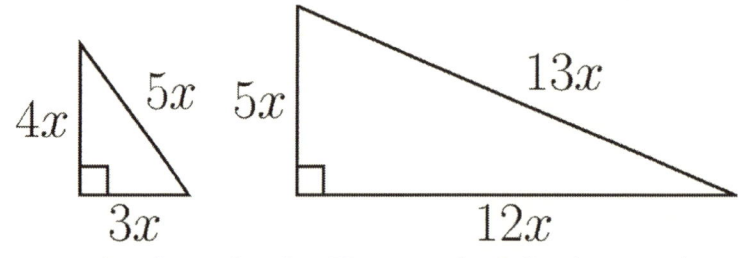

It's not enough just to memorize these triangles. You must **look** for them on the test. Too many students waste time doing calculations, **not** because they don't know these special right triangles but because they don't **notice** them when they appear.

74 |Day 6

Another concept to know is **similarity**, which on the test most often applies to triangles. If 2 triangles are similar, they have the same angles, which gives them the same shape, and their side lengths have the same **ratio**. So if the longest side of a triangle is 3 times its shortest side, this is also true for a similar triangle.

Okay, there's one more, and then on to the drill. As you might know, when 2 parallel lines are intersected by a third line, 8 angles are formed. The "big" angles are all the same, and the "little" ones are all the same, like this:

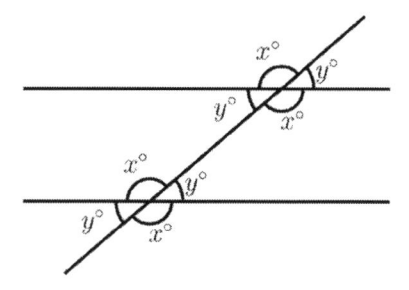

Day 6

Drill

Complete all the questions below.

1.

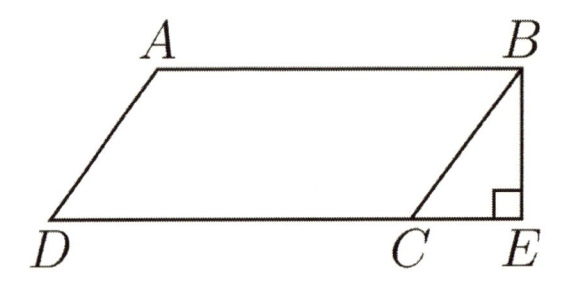

In parallelogram *ABCD* above, the length of side $\overline{AB}$ is 10, and side $\overline{BC}$ is 5. If $\overline{DE}$ is 13, what is the area of parallelogram ABCD?

2.

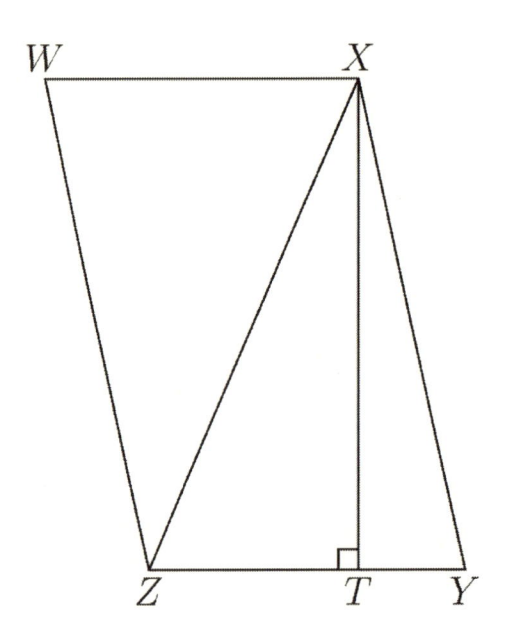

In the figure above, $\overline{XZ}$ is a diagonal of parallelogram WXYZ, and the length of $\overline{ZY}$ is 12. If the area of triangle WXZ is 108, what is the length of $\overline{XT}$?

3.

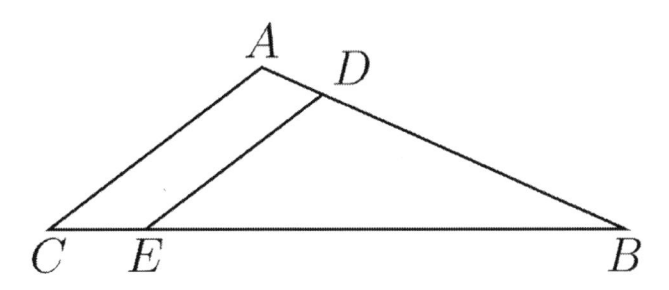

In the figure above, $\overline{AC}$ is 24 inches long and is parallel to $\overline{DE}$, which is 20 inches long. If $\overline{DB}$ is 30 inches long, what is the length of $\overline{AD}$ in inches?

4.

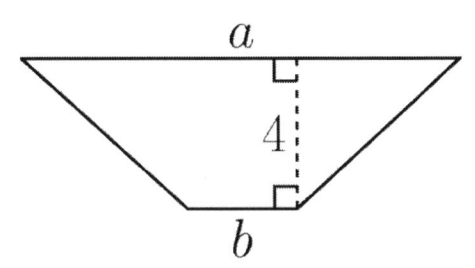

In the trapezoid above, $a = 4b$. If the area of the trapezoid is 30, which of the following is equal to b?

5. What is the area of a circle that has a circumference of 12π?

6.

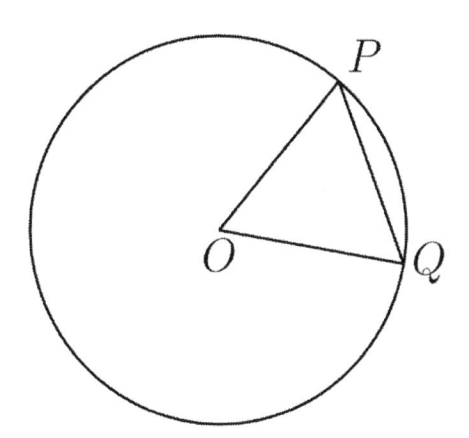

The diameter of circle O above is 18, and the measure of $\angle POQ$ is 60°. What is the length of $\overline{PQ}$?

Drill | 77

7. The radius of a right circular cylinder with a volume of 50π is 10. What is the height of the cylinder?

8. To the nearest integer value, what is the measure of one angle of a regular polygon with 7 sides?

9. If the length of the legs of a triangle are 2.5 and 6, what is the length of the hypotenuse?

10. If the hypotenuse of a triangle is *2x* and one of the legs is *x*, the hypotenuse is how much longer than the other leg?

        A) x
        B) $x(x - \sqrt{3})$
        C) $x(2 - \sqrt{3})$
        D) $x(\sqrt{3} - 1)$

11. What is the area of an isosceles right triangle with a hypotenuse of length 10?

12.

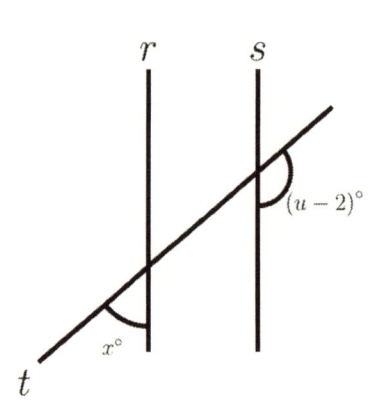

In the figure, line *r* is parallel to line *s*. Which of the following represents *x*?

        A) $u - 2$
        B) $90 + u$
        C) $180 - u$
        D) $180 + u$

Day 6

Drill Explanations

1.

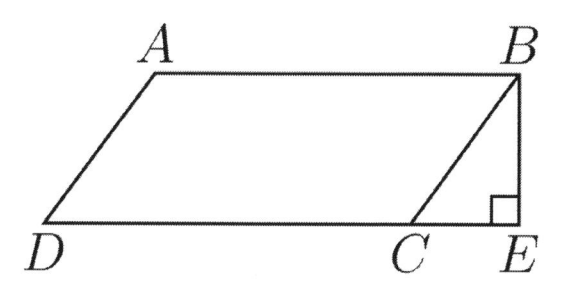

In parallelogram *ABCD* above, the length of side $\overline{AB}$ is 10, and side $\overline{BC}$ is 5. If $\overline{DE}$ is 13, what is the area of the parallelogram?

The area of a parallelogram is *bh* (base x height). We can consider either $\overline{AB}$ or $\overline{DC}$ the base since they're the same. So the base is 10. The height is always at right angles to the base, so the height is $\overline{BE}$. This segment is 1 leg of a right triangle. That's great since we know a lot about them. Since $\overline{DE}$ is 13 and $\overline{DC}$ is 10, the leg $\overline{CE}$ is 3. The hypotenuse of this triangle, $\overline{BC}$, is 5. That means we have a 3 : 4 : 5 triangle, with $\overline{BE}$ being the 4. With a base of 10 and a height of 4, the area is 40.

2.

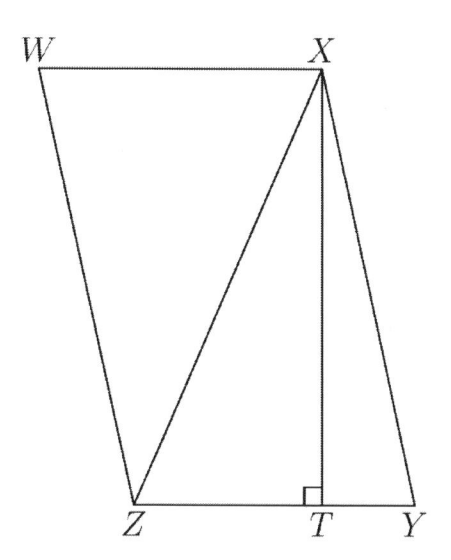

In the figure above, $\overline{XZ}$ is a diagonal of parallelogram $WXYZ$, and the length of $\overline{ZY}$ is 12. If the area of triangle WXZ is 108, what is the length of $\overline{XT}$?

This parallelogram is divided by the diagonal into 2 congruent (identical) triangles. Therefore, the area of triangle ZXY is also 108. The base of this triangle is given as 12. Since XT is the height (h) of this triangle, we can say that $\frac{1}{2}(12h) = 108$, because one-half of base x height is the area of a triangle. This simplifies to $6h = 108$, and $h = 18$.

3.

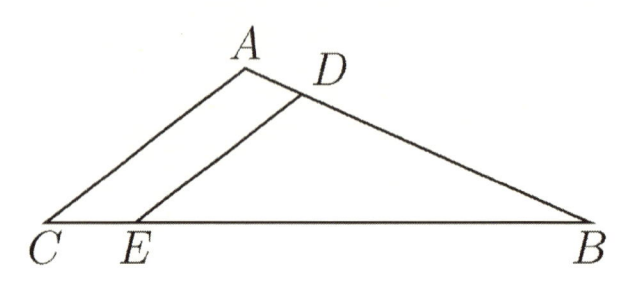

In the figure above, $\overline{AC}$ is 24 inches long and is parallel to $\overline{DE}$, which is 20 inches long. If $\overline{DB}$ is 30 inches long, what is the length of $\overline{AD}$ in inches?

When we see triangles within triangles, *similarity* is almost certainly being tested. The 2 corresponding sides have lengths of 24 and 20, which simplifies to the ratio 6 : 5. If we use x to represent the length of side $\overline{AB}$, we can set up this equation as $\frac{x}{30} = \frac{6}{5}$. With a little cross-multiplication, we get $5x = 180$. And $x = 36$. That's the length of $\overline{AB}$. **Be careful**, because this might show up as an answer choice, but it's a trap. The question asks for $\overline{AD}$, so we must subtract 30 from 36 to get 6.

4.

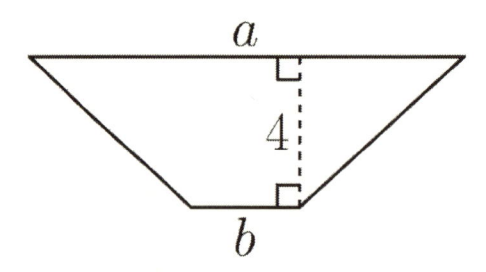

In the trapezoid above, $a = 4b$. If the area of the trapezoid is 30, which of the following is equal to b?

The area of a trapezoid is equal to the average of the bases times the height, so we can set up the following equation: $\frac{a+b}{2} \times 4 = 30$. Since $a = 4b$, we can use substitution to say $\frac{5b}{2} \times 4 = 30$. This simplifies to $10b = 30$, and $b = 3$.

5. What is the area of a circle that has a circumference of 12π?

80 |Day 6

Given a circumference of 12π, we know immediately that the diameter is 12. That means the radius is 6, and *that* means the area is 36π.

6.

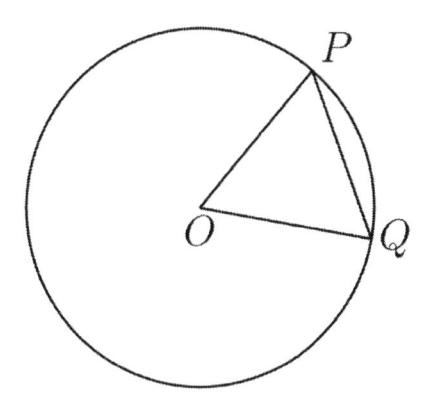

The diameter of circle O above is 18, and the measure of ∠POQ is 60°. What is the length of $\overline{PQ}$?

The segments $\overline{OP}$ and $\overline{PQ}$ are radii (the plural of radius) of the circle, so they must be equal in length. The angles opposite these sides are therefore also equal. Since ∠POQ is 60°, this must be an equilateral triangle, a triangle with equal sides and 3 60-degree angles. (The sum, of course, is 180 degrees.) Since the diameter is 18, the radius is 9, and $\overline{PQ}$ is also 9.

7. The radius of a right circular cylinder with a volume of 50π cubic inches is 10 inches. What is the height of the cylinder in inches?

The formula for the volume of a right circular cylinder, like a glass with straight sides or a length of pipe, is $\pi r^2 h$. We can therefore say $50\pi = \pi(10)^2 h$. That simplifies to $50\pi = 100\pi h$. We can divide each side by π and get $50 = 100h$. That means the height is ½. **Be careful!** It's likely that 2 will show up as a trap choice.

8. To the nearest integer value, what is the measure of 1 angle of a regular polygon with 7 sides?

We'll use the formula $180(n - 2)$ where n is the number of sides. That gives us $180(7 - 2)$, which simplifies to 180 times 5, which is 900. This is the sum of the angles. A regular polygon has equal angles, so we now need to divide this by 7. When we divide 900 by 7, we get a repeating decimal that is closest to the integer 129.

9. If the length of the legs of a right triangle are 2.5 and 6, what is the length of the hypotenuse?

You could use the Pythagorean theorem here, but always be on the lookout for special right triangles. If we double the given values, we get 5 and 12. Those are the ratio values of the legs of a 5 : 12 : 13 triangle, so the hypotenuse is half of 13, which is 6.5.

10. If the hypotenuse of a right triangle is $2x$ and one of the legs is x, the hypotenuse is how much longer than the other leg?

A) x
B) $x(x - \sqrt{3})$
C) $x(2 - \sqrt{3})$
D) $x(\sqrt{3} - 1)$

Right away, we know this is a 30/60 right triangle because the hypotenuse is twice the (shorter) leg. We also have memorized (right?) that the other (longer) leg is equal to the shorter leg times $\sqrt{3}$. That means this leg is $x\sqrt{3}$. We want the *difference* between the hypotenuse and this leg (not how many *times* longer), so that's $2x - x\sqrt{3}$. While this is correct, it's not in the form given in the choices, so let's factor. We can factor x from each term, giving us **choice C**, $x(2 - \sqrt{3})$.

11. What is the area of an isosceles right triangle with a hypotenuse of length 10?

This is one of the *special* right triangles. Besides the right angle, it has 2 45-degree angles. The hypotenuse is equal to the length of one of the legs *multiplied* by $\sqrt{2}$. That means one of the legs is equal to 10 *divided* by $\sqrt{2}$. We could just use $\frac{10}{\sqrt{2}}$ as the length of a leg and say $\frac{(\frac{10}{\sqrt{2}})(\frac{10}{\sqrt{2}})}{2} = area$. That's not as bad as it might look. It simplifies to $\frac{50}{2}$, and we get 25. (When we simplify the numerator of that complicated-looking fraction, we multiply across the top and get 100 and then multiply across the bottom to get 2. Remember that $\sqrt{2}$ times $\sqrt{2}$ equal 2, and $\sqrt{17}$ times $\sqrt{17}$ equals 17, and so on. The numerator then simplifies to 50, and when we divide that by the denominator, we get 25.)

You might have decided to take $\frac{10}{\sqrt{2}}$, which is the length of a leg, and *rationalize the denominator*, something we often do when there's a root down there. To do that, we multiply in the following way: $\frac{10}{\sqrt{2}} \times \frac{\sqrt{2}}{\sqrt{2}} = \frac{10\sqrt{2}}{2} = 5\sqrt{2}$. That's fine, and if we use this form of the leg to get the hypotenuse, we, of course, get the same answer: $\frac{(5\sqrt{2})(5\sqrt{2})}{2} = \frac{50}{2} = 25$.

12.

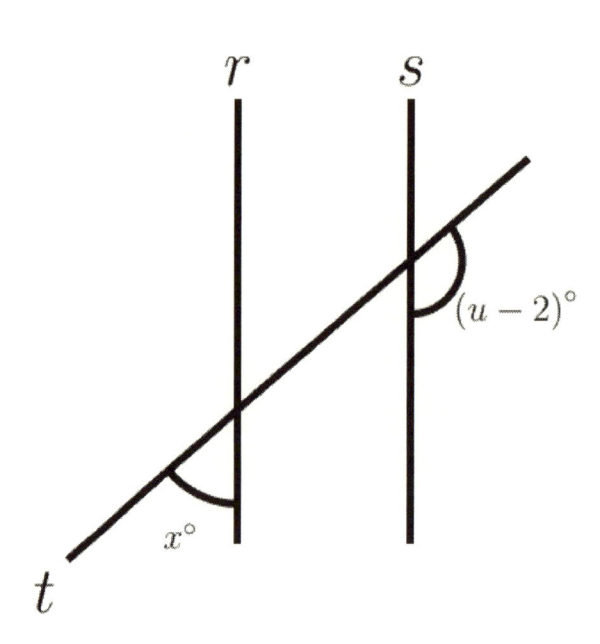

In the figure, line *r* is parallel to line *s*. Which of the following represents *x*?

A) $u - 2$
B) $90 + u$
C) $182 - u$
D) $180 + u$

Line *t* creates 8 angles by intersecting the other 2 lines, and since those other 2 lines are parallel, all the big angles are equal, and all the little ones are equal. Furthermore, the sum of a little and a big angle is 180 degrees. They are supplementary—they form a straight line, which is 180 degrees.

We can therefore create the equation $x + (u - 2) = 180$. To isolate *x*, we can subtract $u - 2$ from both sides. That gives us $x = 180 - (u - 2)$. This simplified to $180 - u + 2$ or $182 - u$. This is choice C.

Drill Explanations | 83

Day 7

Quiz 1

After reading this material, reviewing days 1–5, and completing Quiz 1, sign and check the box for Day 7.

Today is a day for review, but make no mistake about it, it's **hugely** important that you do this. Too many students are exposed to math content, understand the math content, answer some questions about that content, and **still mess up on that content** on test day. Knowing isn't enough; **using** what you know is what this is all about.

To get to this stage, you need to clear a path in your brain so when you need to find the circumference of a circle, factor a tricky quadratic, or simplify a fraction with a negative exponent in the denominator, your proper neurons get activated and you move confidently and efficiently down that path until you solve the problem. You get to that stage through **review**.

It's rare that we see anything surprising on the SAT. What you *will* see are variations on the material in this course. You need to be able to recognize these variations for what they are and not freak out because a question looks so weird and hard. You get to that non-freaking-out stage through **review.**

We've covered a good deal by now. For today, you'll complete the quiz below, but it's an **excellent idea** to flip back to the earlier days and find the material that at the time you thought, "Yeah, I pretty much get that" or "Yeah, I got that one, but it took me a little too long." That way you can **master** that idea, formula, or whatever.

For example, on the day of the test, there's a good chance that a question is going to include a 30/60 triangle.

- ☐ Some students will have no idea what to do, and they'll get it wrong.
- ☐ Some students will not know about the 30/60, but they'll get the question right in about 1.5 minutes.
- ☐ Some students *will* know about the 30/60 and will not realize that they're looking at it, and they'll also get the question right in about 1.5 minutes.
- ☐ Some students will know about the 30/60—sorta kinda. Eventually, after doubting themselves, they'll recall the $1:\sqrt{3}:2$ ratio of these triangles, and (if they don't get careless) they'll get it right in about 1 minute and 10 seconds. And then—
- ☐ Some students—not that many—recognize this triangle, stroll down the path in their brains that leads right to the $1:\sqrt{3}:2$ ratio, and solve the problem in 32 seconds. **Wanna be that student?**

Once you've reviewed the past days' lessons, take the following quiz. Strive for—that's right—efficiency!

Note: The geometry from Day 6 will be on the next quiz, not this one.

Quiz 1

1. $\dfrac{x}{12} + \dfrac{5}{4y} =$

2. $\dfrac{a+2b}{3c} - \dfrac{ab}{5c^2 d} =$

3. $\dfrac{x}{7y} \times \dfrac{ay}{4} =$

4. $\dfrac{2t-a}{t} \times \dfrac{t}{v} \times \dfrac{t}{a} =$

5. $\dfrac{\frac{3a}{2}}{7} =$

6. $\dfrac{\frac{a}{2x}}{\frac{xy}{a+1}} =$

7. $\dfrac{a}{3+a} = \dfrac{5}{2}, a =$

8. $-\dfrac{b}{6} = \dfrac{3b+1}{4}, b =$

9. Split the following into 2 fractions. Simplify if possible. $\dfrac{3xy^2 - 16z}{2yz}$

10. Along a particular street, the ratio of cars to bicycles on Mondays between 10 and 11 in the morning is estimated at 29 : 3. If, during this time one Monday, there are a total of 256 cars and bicycles, what is the best estimate of how many more cars than bicycles are on the road?

11. What is 120% as a fraction and as a decimal?

12. What is 5/6 as a percentage and a decimal?

13. What is 0.012 as a percentage and a fraction?

14. 5% of 30 =

15. 9% of 10 =

16. 5 is what percentage of 25?

17. 0.15 is what percentage of 6?

18. 2 is what percentage less than 3?

19. 3 is what percentage greater than 2?

20. $7^x \times 7^{x+1} =$

21. $\left(x^{-\frac{1}{2}}\right)\left(x^{\frac{5}{2}}\right) =$

22. $\dfrac{(ab)^{10}}{(ab)^3} =$

23. $\dfrac{r^{2a+1}}{r^{a-1}} =$

24. $\left(\dfrac{3}{2}\right)^{-3} =$

25. $16^{\frac{5}{4}} =$

26. $\sqrt{\dfrac{25}{36}} =$

27. $\sqrt{98} =$

28. $x - 14 = 5x, 4x =$

29. $\dfrac{a - bc}{3} = ab, 3ab + bc - a =$

30. $st - 5st = 4s, t =$

For the next two questions, isolate r and then y in 1 inequality, or possibly 2.

31. $rs - r > s$, and $s < 0$

32. $|y| > 1$

33. $x^2 - 2x = 99$, $x =$

34. Factor $r^2 - s^2$

35. Solve $5x^2 + 12x + 4 = 0$ by grouping.

36. Express $x^2 - 6x + 34 = 0$ in the form $a(x - h)^2 + k$. Hint: a is 1.

37. What is the quadratic formula?

38. How many minutes and seconds did it take a car moving at an average speed of 40 mph to travel 35 miles?

39. The average age of 20 people in a room is 32. If the youngest person in the room is 18 and the oldest person is 50, what is the sum of the ages of the other 18 people?

40. Last year, a furniture store ordered 532 items—all of them doors and locks—from a manufacturer. Each door costs the store $120, and each lock costs $20. The store paid $40,640 for all these items. This year, the store will order 7 times the number of doors and twice as many locks as were ordered last year. How many of these items were ordered this year?

41. The managers of a community pool use the equation $n = (t - 85)^2 + 40$ to approximate how many people (n) will use the pool in a day based on an average temperature (t) between 70 and 100 degrees. Based on this equation, at what temperature will the least number of people use the pool?

Quiz 1 Explanations

1. $\dfrac{x}{12} + \dfrac{5}{4y} = \dfrac{xy+15}{12y}$ 2. $\dfrac{a+2b}{3c} - \dfrac{ab}{5c^2 d} = \dfrac{5cd(a+2b)}{15c^2 d} - \dfrac{3ab}{15c^2 d} = \dfrac{5acd+10bcd-3ab}{15c^2 d}$

3. $\dfrac{x}{7y} \times \dfrac{ay}{4} = \dfrac{ax}{28}$ 4. $\dfrac{2t-a}{t} \times \dfrac{t}{v} \times \dfrac{t}{a} = \dfrac{2t-a}{\not{t}} \times \dfrac{\not{t}}{v} \times \dfrac{t}{a} = \dfrac{2t^2 - at}{av}$ 5. $\dfrac{\frac{3a}{2}}{7} = \dfrac{3a}{14}$

6. $\dfrac{\frac{a}{2x}}{\frac{xy}{a+1}} = \dfrac{a}{2x} \times \dfrac{a+1}{xy} = \dfrac{a^2+a}{2x^2 y}$ 7. $\dfrac{a}{3+a} = \dfrac{5}{2}$ $2a = 15 + 5a$ $3a = -15$ $a = -5$

8. $-\dfrac{b}{6} = \dfrac{3b+1}{4}$ $-4b = 18b + 6$ $22b = -6$ $b = -\dfrac{3}{11}$

9. Split the following into 2 fractions. Simplify if possible. $\dfrac{3xy^2 - 16z}{2yz} = \dfrac{3xy^2}{2yz} - \dfrac{16z}{2yz} = \dfrac{3xy}{2z} - \dfrac{8}{y}$

10. Along a particular street, the ratio of cars to bicycles on Mondays between 10 and 11 in the morning is estimated at 29 : 3. If, during this time one Monday, there are a total of 256 cars and bicycles, what is the best estimate of how many more cars than bicycles are on the road?

Two equations can be set up: $\dfrac{c}{b} = \dfrac{29}{3}$ and $c + b = 256$. After cross-multiplying the first equation and then using substitution, we get $c = 232$ and $b = 24$. That means there are 208 more cars than bicycles.

11. What is 120% as a fraction and as a decimal? $\dfrac{12}{10} = \dfrac{6}{5} = 1.2$

12. What is 5/6 as a percentage and a decimal? $0.8\overline{3} = 83\tfrac{1}{3}\%$

13. What is 0.012 as a percentage and a fraction? $1.2\% = \dfrac{12}{1000} = \dfrac{3}{250}$

14. 5% of 30 = 1.5   15. 9% of 10 = 0.9   16. 5 is what percentage of 25? $\dfrac{5}{25} = 20\%$

17. 0.15 is what percentage of 6? $\dfrac{0.15}{6} = 2.5\%$ 18. 2 is what percentage less than 3? $\dfrac{1}{3} = 33\tfrac{1}{3}\%$

19. 3 is what percentage greater than 2? $\dfrac{1}{2} = 50\%$

20. $7^x \times 7^{x+1} = 7^{2x+1}$ 21. $\left(x^{-\frac{1}{2}}\right)\left(x^{\frac{5}{2}}\right) = x^2$ 22. $\dfrac{(ab)^{10}}{(ab)^3} = (ab)^7$ 23. $\dfrac{r^{2a+1}}{r^{a-1}} = r^{a+2}$

24. $\left(\dfrac{3}{2}\right)^{-3} = \left(\dfrac{2}{3}\right)^3 = \dfrac{8}{27}$ 25. $16^{\frac{5}{4}} = 32$ 26. $\sqrt{\dfrac{25}{36}} = \dfrac{5}{6}$ 27. $\sqrt{98} = 7\sqrt{2}$

28. $x - 14 = 5x$, $4x = -14$ 29. $\dfrac{a-bc}{3} = ab$, $3ab + bc - a = 0$

Quiz 1 | 87

30. $st - 5st = 4s - 4st = 4s - 4t = 4$ $t = -1$

For the next 2 questions, isolate r and then y in 1 inequality, or possibly 2.

31. $rs - r > s$, and $s < 0$ $r(s-1) > s$ $r < \frac{s}{s-1}$
Note: The sign gets flipped because we divided by a negative. (Since s is negative, so is $s - 1$.)

32. $|y| > 1$ $y > 1$ and $y < -1$

33. $x^2 - 2x = 99$, $x = -9$ and $x = 11$ Note: Set equal to 0, and factor or use the on-screen calculator graph to find the 0's where the parabola crosses the x-axis.

34. Factor $r^2 - s^2$ $(r+s)(r-s)$

35. Solve $5x^2 + 12x + 4 = 0$ by grouping $5x^2 + 10x + 2x + 4 = 0$ $5x(x+2) + 2(x+2) = 0$ $(5x+2)(x+2) = 0$ $x = -\frac{2}{5}$ and $x = -2$

36. Express $x^2 - 6x + 34 = 0$ in the form $a(x-h)^2 + k$. Hint: a is 1. $(x-3)^2 + 25$
Note: This can be solved by completing the square (see Day 4) or by using the OSC. If you graph the quadratic, you'll see that the vertex is at $(3, 25)$.

37. What is the quadratic formula? $x = \frac{-b \pm \sqrt{b^2 - 4ac}}{2a}$

38. How many minutes and seconds did it take a car moving at an average speed of 40 mph to travel 35 miles?
Distance over rate $= \frac{35}{40} = \frac{7}{8}$. This is 7/8 of an hour. This is 52.5 or 52 minutes and 30 seconds.

39. The average age of 20 people in a room is 32. If the youngest person in the room is 18 and the oldest person is 50, what is the sum of the ages of the other 18 people?

The sum of all the ages is 20 x 32, which is 640. Since the sum of the youngest and oldest person is 68, the other ages total to 640 – 68, which is 572.

40. Last year, a furniture store ordered 532 items—all of them doors and locks—from a manufacturer. Each door costs the store $120, and each lock costs $20. The store paid $40,640 for all these items. This year, the store will order 7 times the number of doors and twice as many locks as were ordered last year. How many of these items were ordered this year?

We can set up the equations $d + l = 532$ and $120d + 20l = \$40,640$. In these equations, d and l represent the number of doors and the number of locks ordered last year. Before using substitution, you could simplify the second equation by dividing by 10 and then by 2, giving you $6d + l = 2032$. Rather than solving for each unknown, it's a simple matter to add the first equation to this simplified second equation to get what we're looking for: 7 times the number of doors and twice the number of locks. Adding the equations gives us $7d + 2l = 532 + 2032 = 2564$.

41. The managers of a community pool use the equation $n = (t - 85)^2 + 40$ to approximate how many people (n) will use the pool in a day based on an average temperature (t) between 70 and 100 degrees. Based on this equation, at what temperature will the least number of people use the pool?

This quadratic is in vertex form. That means the vertex is at (85, 40). This is an upward opening parabola, so the least number of people (40) will use the pool when the temperature is 85 degrees.

Day 8

Rest, Review, Quiz 1 Explanations

After reading this material and completing the drill at the end, sign and check the box for Day 8.

Coordinate Geometry: Know How to Calculate Distance, Slope, and Midpoint

With coordinate geometry, we locate shapes in the *xy*-coordinate plane. Let's first plot a few points.

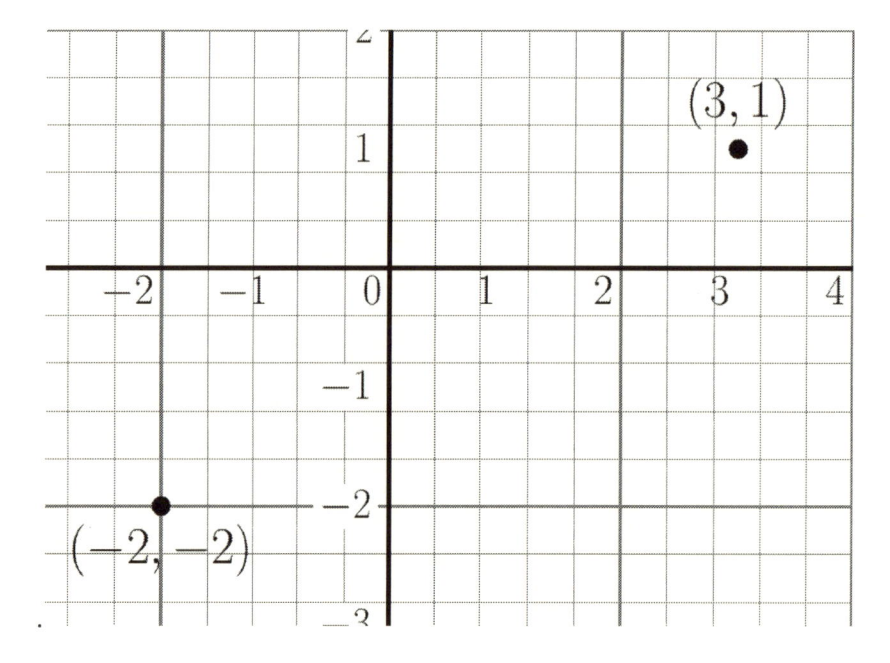

Given the points (3, 1) and (-2, -2), we might be asked several things. Let's start with this: "What's the distance between these points?" Since we can't simply count this distance, we'll **triangulate**. That means we'll connect the 2 points and make that line segment into the hypotenuse of a right triangle. We can now easily determine the length of the 2 legs: 3 and 5.

90 |Day 8

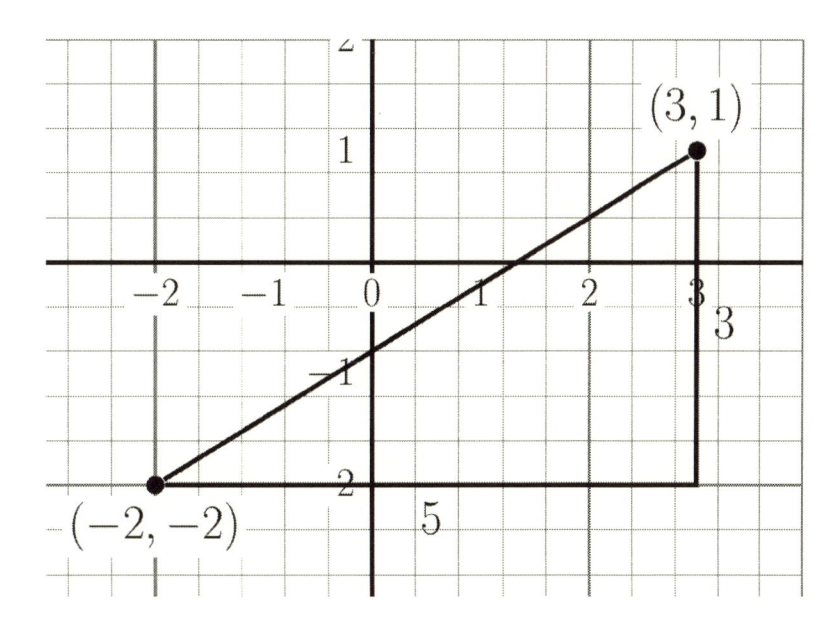

Great! Once we know any 2 sides of a right triangle, we can always calculate the 3rd. *First*, we want to see if we have a special right triangle—and we don't. It's not a 45/45 or a 30/60 or a 3 : 4 : 5 or a 5 : 12 : 13. (You haven't memorized those? Stop everything, and do so **now**.) So we'll use the Pythagorean theorem, $3^2 + 5^2 = c^2$ so $c^2 = 34$. This means that c, the hypotenuse, which is what we want for this question, is $\sqrt{34}$. That is an irrational number, so we can't do anything else with it. The distance between the two points is $\sqrt{34}$.

Something else you might be asked in this situation is the **slope** of the line segment connecting the 2 points. As you probably know, to calculate the slope, we divide the difference of the y-coordinates of 2 points by the difference of the x-coordinates of the 2 points. This is often called "rise over run." With these 2 points, we have $\frac{1-(-2)}{3-(-2)} = \frac{3}{5}$. (Actually, we already did the work, didn't we? When we were calculating distance, we determined that the rise was 3 and the run was 5.)

Once we get more into word problems, it's often useful to think of slope as being a **rate of change**. How quickly do the y values increase (or decrease) as the x values increase? For example, if the x-axis shows "**dollars earned**" increasing from left to right and the y-axis shows "**hours worked**" increasing from bottom to top, a line with a positive slope drawn in the plane would show the rate someone's earnings increased per hour.

Something else to remember about slope is that lines that are perpendicular to each other have slopes that are **negative reciprocals**. Since our line (segment) above has a slope of $\frac{3}{5}$, a line perpendicular to it would have a slope of $-\frac{5}{3}$.

Here's one more thing you might be asked about points (3, 1) and (-2, -2). What is the **midpoint**? To get the midpoint, we'll use that most useful of formulas, the **average** formula. We just use it twice, once to get the x-coordinate and once to get the y-coordinate. That gives us $\frac{3+(-2)}{2}, \frac{1+(-2)}{2}$, which is equal to $(\frac{1}{2}, -\frac{1}{2})$.

Rest, Review, Quiz 1 Explanations | 91

Lines: Know the Equation $y = mx + b$

In this equation, as you probably know, the m value is the slope of the line, and b is the **y-intercept**, which means the y value where the line crosses the y-axis. At this point, $x = 0$. Lines also have **x-intercepts**. This is naturally the point on a line where $y = 0$.

Lines are not the only shapes that are graphed onto the coordinate plane. You'll also see the following.

Circles: Know the Equation $(x - h)^2 + (y - k)^2 = r^2$

In this equation, (h, k) are the coordinates of the center of the circle, and r is the radius. For example, if we graph $(x - 2)^2 + (y + 3)^2 = 16$, we get the following:

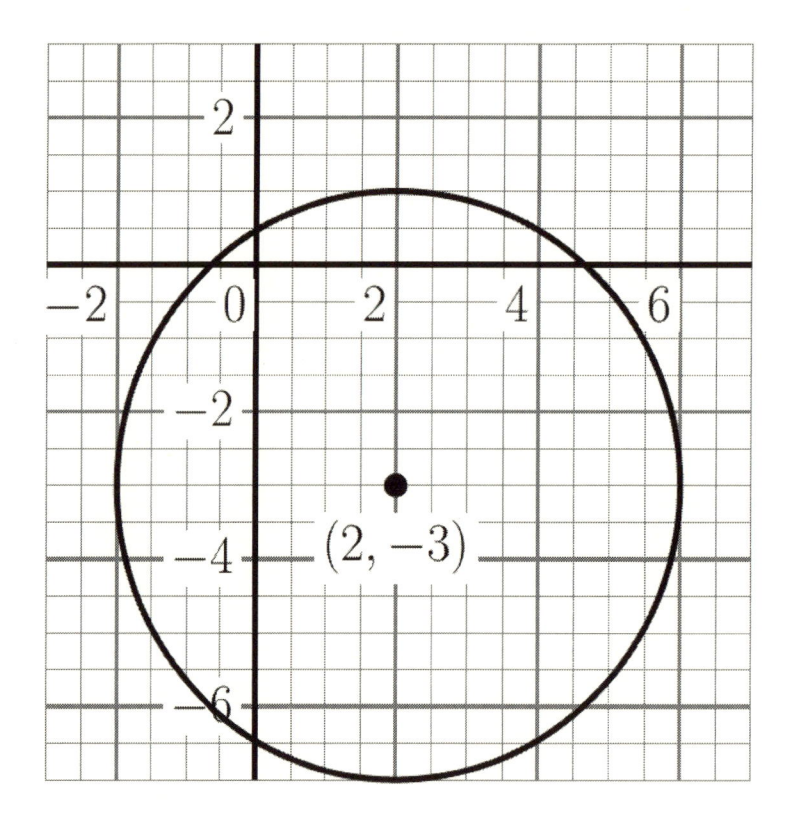

If this equation looks at all familiar, it is derived from the Pythagorean theorem, and that could help you remember it. *Something squared* plus *something squared* equals *something squared*. If you don't see any h or k value in the equation, as in $x^2 + y^2 = 49$, then the circle is centered at (0, 0).

And then we have parabolas.

Parabolas: We're Back to Quadratics

We discussed quadratics a fair amount on Day 4, including how they appear as parabolas when graphed. Remember that they are typically expressed as $ax^2 + bx + c$ or as $a(x - h)^2 + k$, which is known as the *vertex form*, since (h, k) are the coordinates of the vertex. In both cases, if a is positive, the parabola opens upward, and if a is negative, it opens downward. Also, if a is positive, increasing it **narrows** the parabola, and decreasing it **widens** it. The opposite of this (or the upside-down of this) applies to negative a values. Here are some examples.

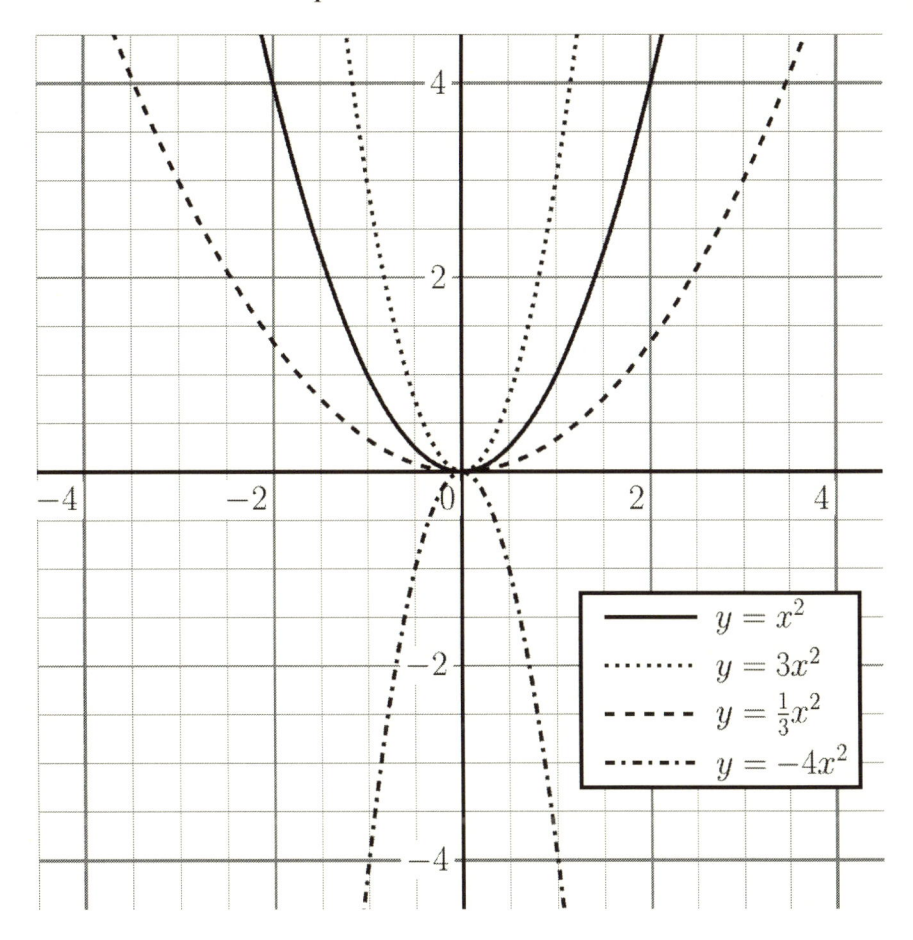

Function Notation: The Use of $y = f(x)$

We'll talk more about function notation soon, but it's worth pointing out now that we frequently see graphs on the test where the corresponding equation looks like, for example, $f(x) = 3x - 1$ or $f(x) = x^2 + 7$. In these cases, the ***f* of *x*** (that's how you say it) is the same as the y value for any point. A question might even make this clear by saying that $y = f(x)$. In the linear equation just mentioned, $f(x) = 3x - 1$, we know that on that line, when x is 4, y is 11. We could say that $f(4) = 11$. If more than 1 graph appears in a problem, you might see $g(x)$, $h(x)$, and so on.

Rest, Review, Quiz 1 Explanations | 93

Day 8

Drill

1. What is the distance in the *xy*-coordinate plane between the points $(-3,-2)$ and $(6,2)$?

2. Line *l* is perpendicular to a line that goes through points $(1,5)$ and $(3,0)$. What is the slope of line *l*?

3. Point *B* is the midpoint of points *A* and *C*. If the coordinates of *B* and *C* are $(7,-2)$ and $(11,4)$, respectively, what are the coordinates of *A*?

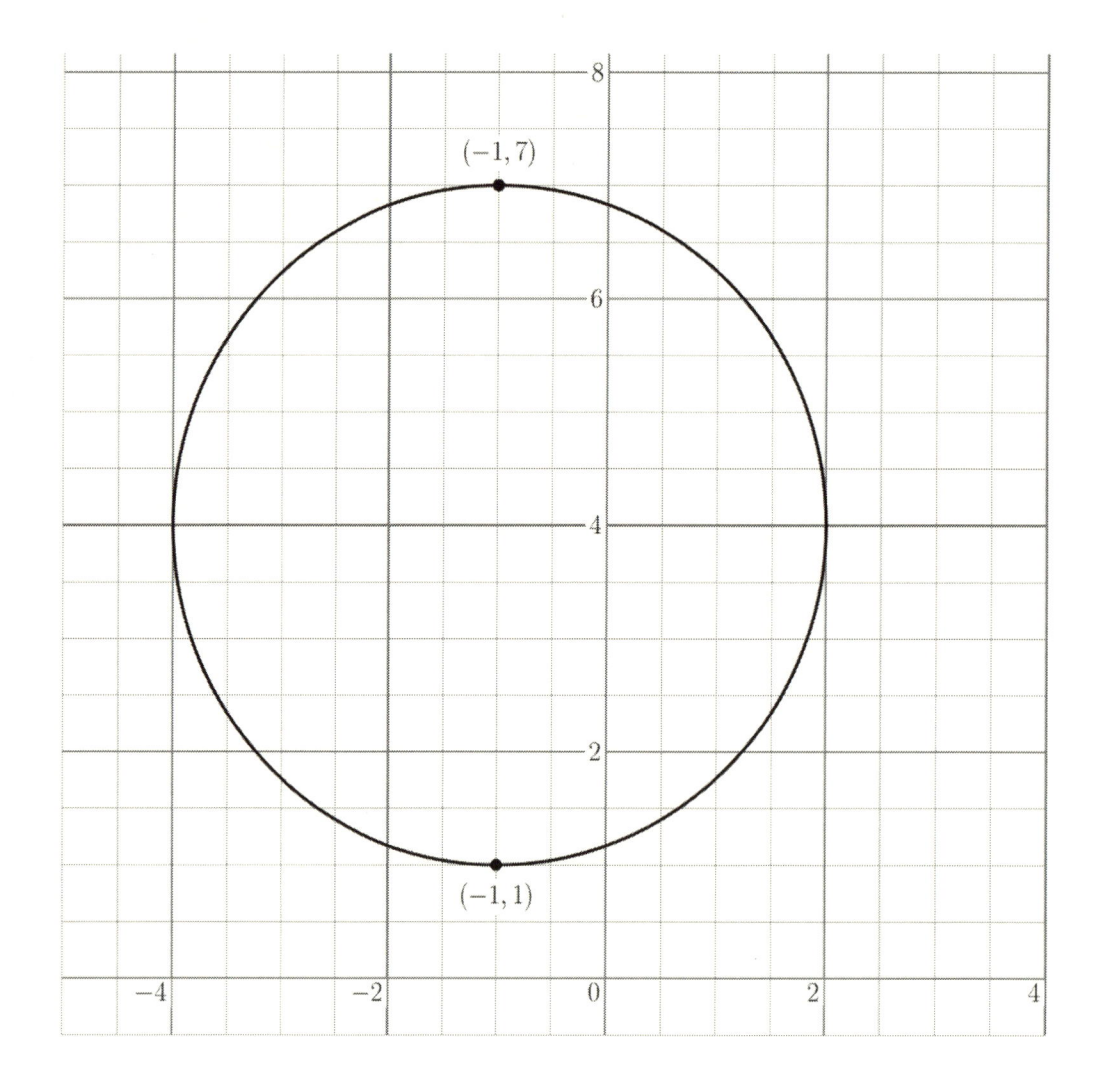

4. What is the equation that describes the circle shown above?

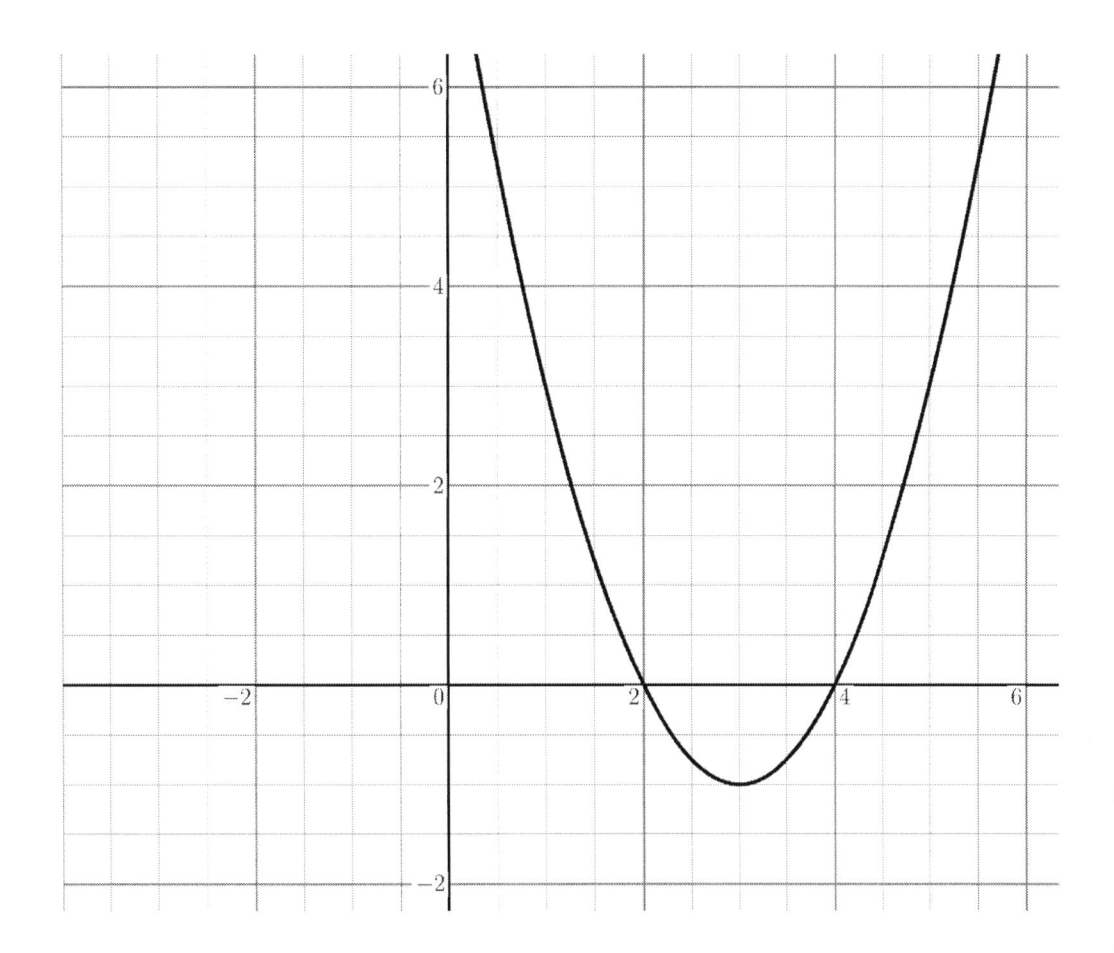

5. If the parabola shown above is shifted up and to the left and no other changes are made, which of the following equations could describe the resulting parabola?

A) $y = (x - 2)^2 - 3$
B) $y = (x - 2)^2$
C) $y = (x - 4)^2$
D) $y = -(x - 2)^2$

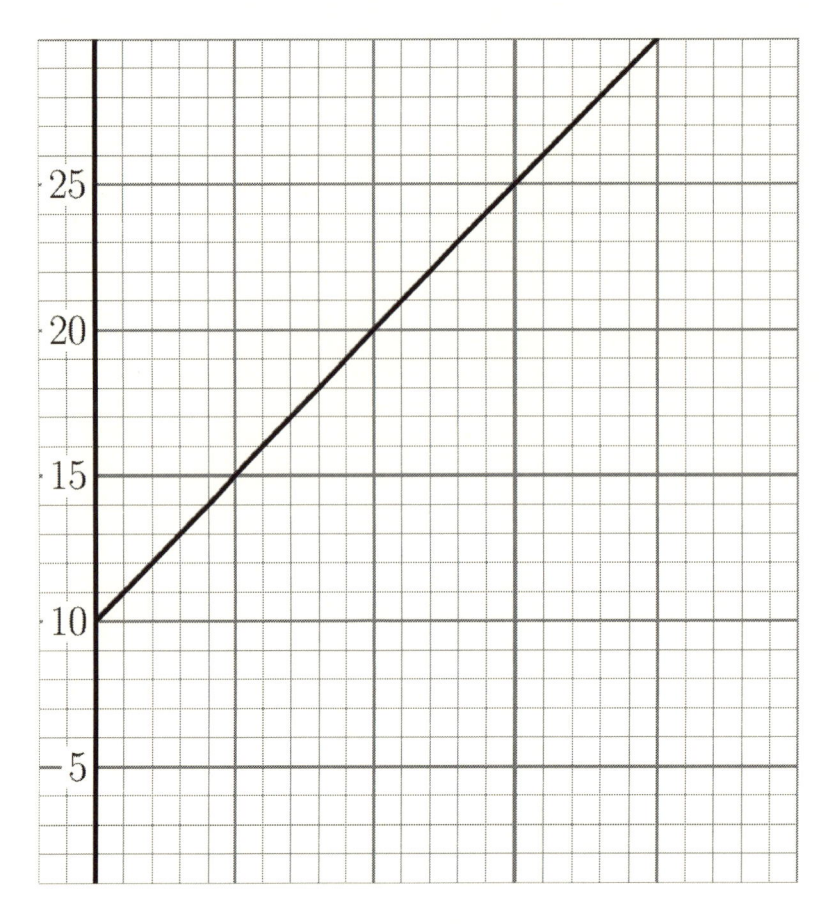

6. The graph above represents the number of people (y) who attended a convention x minutes after the official opening time of the convention. What is the best interpretation of the slope of the graph?

A) The number of people who attended the official time
B) The rate at which the number of people who attended changed per minute
C) The total number of people who attended 10 minutes after the official opening time
D) The increase in the number of people who attended each hour

7. What is the circumference of a circle defined by the equation $(x+9)^2 + (y-9)^2 = 1$?

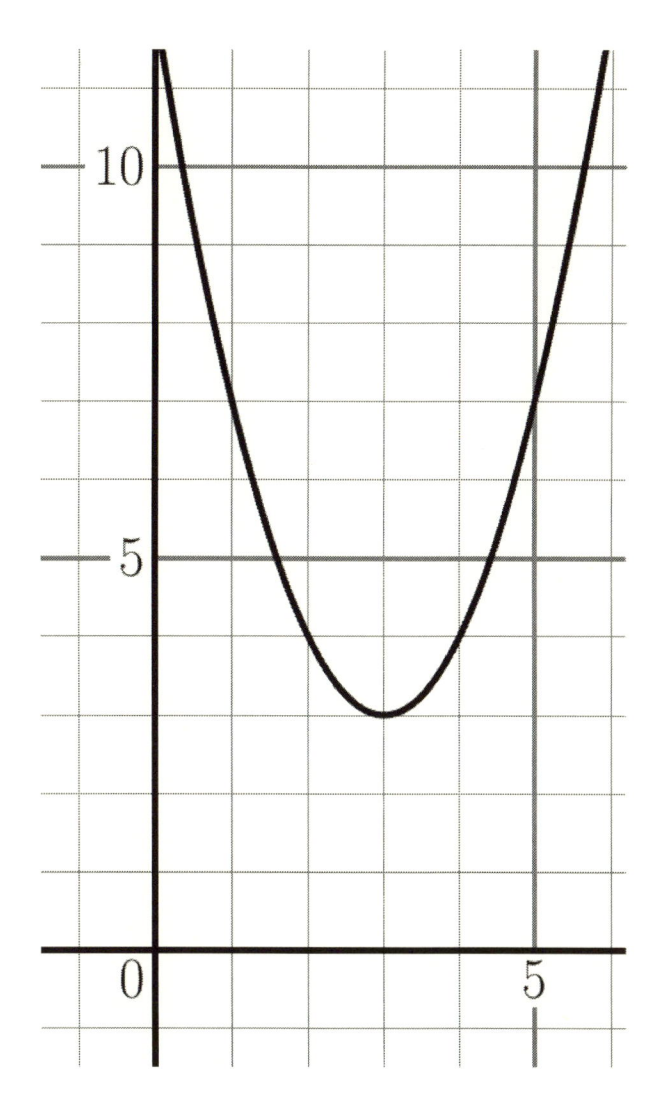

8. Which of the following could be the equation of the graph shown above?

A) $f(x) = x^2 - 6x + 12$
B) $f(x) = x^2 + 6x + 12$
C) $f(x) = x^2 - 6x - 12$
D) $f(x) = -x^2 - 6x + 12$

9. A researcher uses the equation $t = 3(m - 4)^2 + 21$ to model the temperature (t) of a solution m minutes after the beginning of an experiment. Which of the following is the best interpretation of the vertex of the graph of this equation?

A) The maximum temperature of 4 degrees was reached 21 minutes after the beginning of the experiment.
B) The minimum temperature of 4 degrees was reached 21 minutes after the beginning of the experiment.
C) The maximum temperature of 21 degrees was reached 4 minutes after the beginning of the experiment.
D) The minimum temperature of 21 degrees was reached 4 minutes after the beginning of the experiment.

10. What is the x-intercept of the graph of $y = -5x - 6$?

11. Line l passes through points $(1,4)$ and $(-2,1)$. At what point or points, if any, does the graph of $f(x) = x^2 - 9$ intersect line l?

12. Line l is perpendicular to a line that goes through points $(1,5)$ and $(3,0)$. What is the slope of line l?

13. Point B is the midpoint of points A and C. If the coordinates of B and C are $(7,-2)$ and $(11,4)$, respectively, what are the coordinates of A?

Day 8

Drill Explanations

1. What is the distance in the xy-coordinate plane between points $(-3,-2)$ and $(6,2)$?

If you know the distance formula well, that's fine, but here's the graphic approach that uses triangulation.

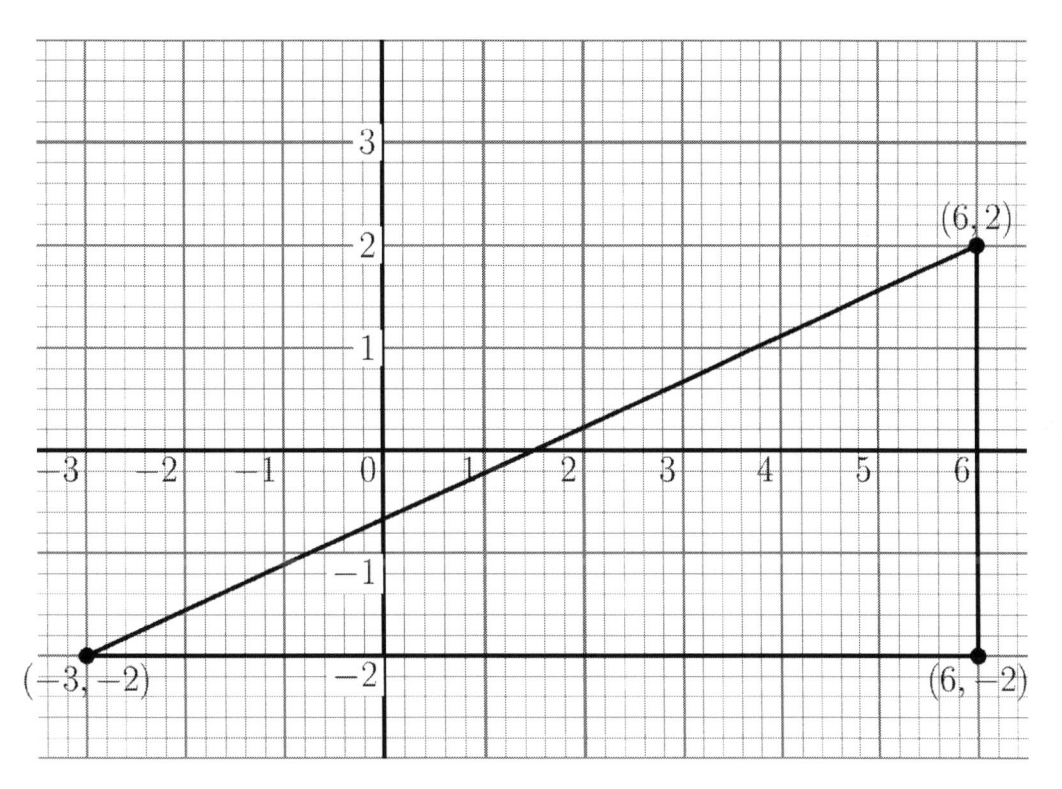

We can now easily determine the lengths of the legs: 9 and 4. This is not a special right triangle, so by using the Pythagorean theorem, we get $9^2 + 4^2 = c^2$, which means that $c^2 = 97$ and $c = \sqrt{97}$.

2. Line l is perpendicular to a line that goes through points $(1,5)$ and $(3,0)$. What is the slope of line l?

The slope of the line that goes through those 2 points is $-\frac{5}{2}$. Line l therefore has a slope of $\frac{2}{5}$.

3. Point B is the midpoint of points A and C. If the coordinates of B and C are $(7,-2)$ and $(11,4)$, respectively, what are the coordinates of A?

We'll call the coordinates of A simply x and y. The x-coordinate of the midpoint (Point B) is 7, so this must be the average of the x-coordinates of the endpoints: 11 and x. Therefore, $\frac{11+x}{2} = 7$. This simplifies to $11 + x = 14$ and $x = 3$. Now we do this again for the y-coordinates. The midpoint has a y-coordinate of -2, so we can say $\frac{4+y}{2} = -2$. This gives us $4 + y = -4$ and $y = -8$. That means the coordinates of point A are $(3, -8)$.

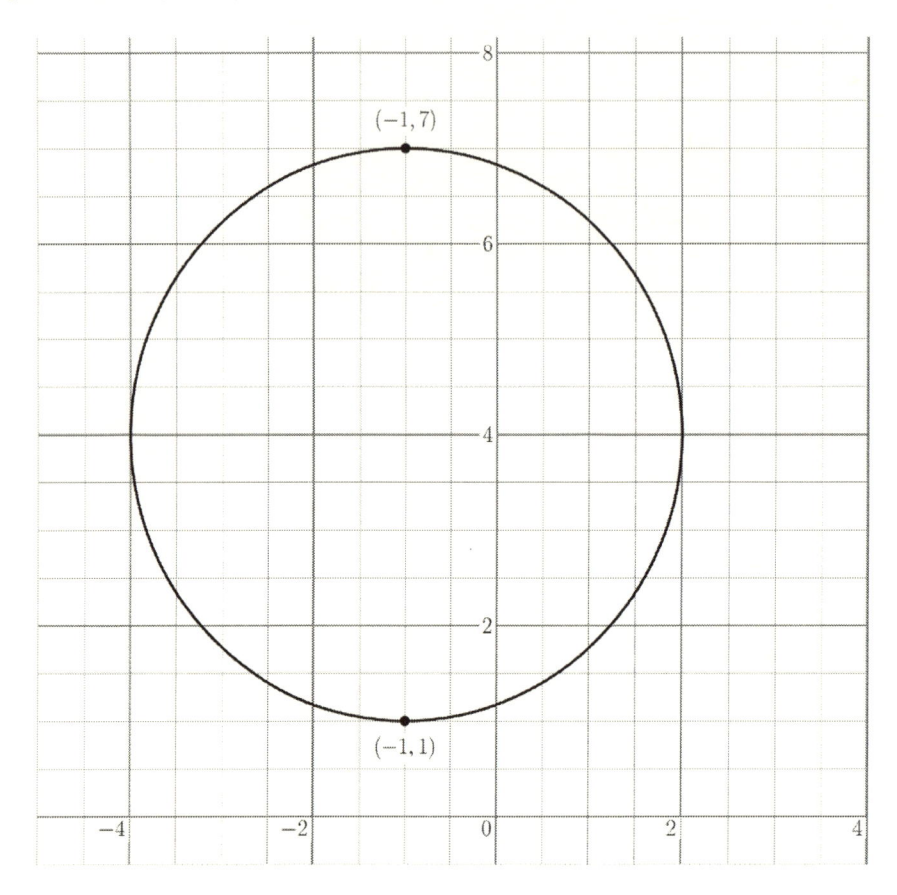

4. What is the equation of the circle shown above?

The equation that describes a circle is $(x - h)^2 + (y - k)^2 = r^2$. The center of the circle is midway between the 2 given points in the diagram. The x coordinate of the center point is the same as it is for those 2 points: **−1**. The y-coordinate for the center point is midway between the 2 given points. Since the distance from 1 to 7 is 6, **the radius is 3**, and the midpoint is 3 "up" from 1, which is **4**. Using these values, we get $(x + 1)^2 + (y - 4)^2 = 9$.

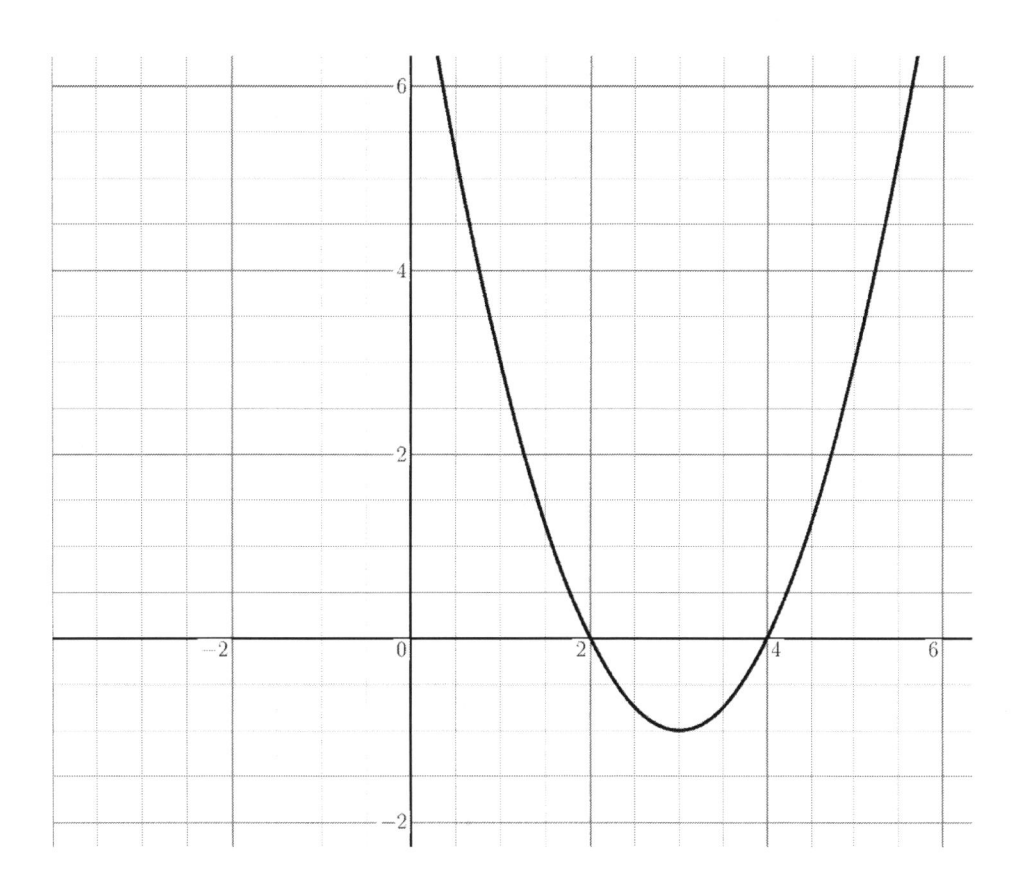

5. If the parabola shown above is shifted up and to the left and no other changes are made, which of the following equations could describe the resulting parabola?

A) $y = (x - 2)^2 - 3$
B) $y = (x - 2)^2$
C) $y = (x - 4)^2$
D) $y = -(x - 2)^2$

The answer choices are quadratics in vertex form: $a(x - h)^2 + k$. The parabola is upward opening and has a vertex that appears to be $(3, -1)$, though we can't say that with absolute certainty. However, shifting it up would **increase** the y-coordinate (which is k in the vertex form) and would not result in a y-coordinate of -3, so we can get rid of choice A. Shifting the parabola to the left would give us a **smaller** x-coordinate (which is h in the vertex form), so we can eliminate choice C. Since this parabola opens upward, we can also eliminate choice D, which, because of the negative 1 implied by the minus sign, would give us a downward opening parabola. That leaves us with **choice B**. This would give us a parabola with a vertex at $(2, 0)$.

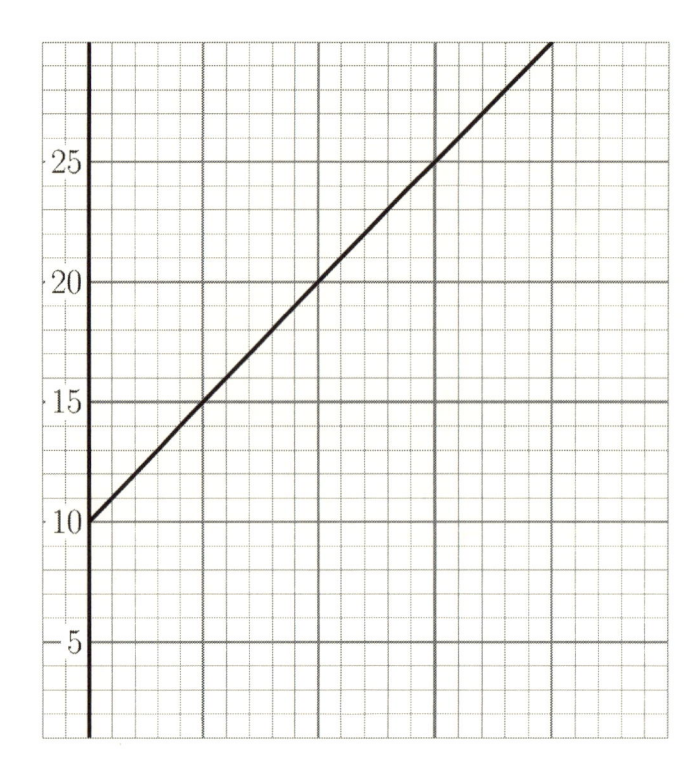

6. The graph above represents the number of people (*y*) who attended a convention *x* minutes after the official opening time of the convention. What is the best interpretation of the slope of the graph?

A) The number of people who attended at the official time
B) The rate at which the number of people who attended changed per minute
C) The total number of people who attended 10 minutes after the official opening time
D) The increase in the number of people who attended each hour

Slope represents **rate of change**. In this case, as the minutes increase (as time goes by), the attendance increases, giving us a positive slope. For example, it appears that every 5 minutes, 5 more people were in attendance. That is a rate of 1 person per minute. This rate change is described in choice B.

7. What is the circumference of a circle defined by the equation $(x + 9)^2 + (y - 9)^2 = 1$?

If you know your equation for the graph of a circle—and you should—you'll know that the radius squared in this case is 1. That means the radius is 1, the diameter is 2, and the circumference is 2π.

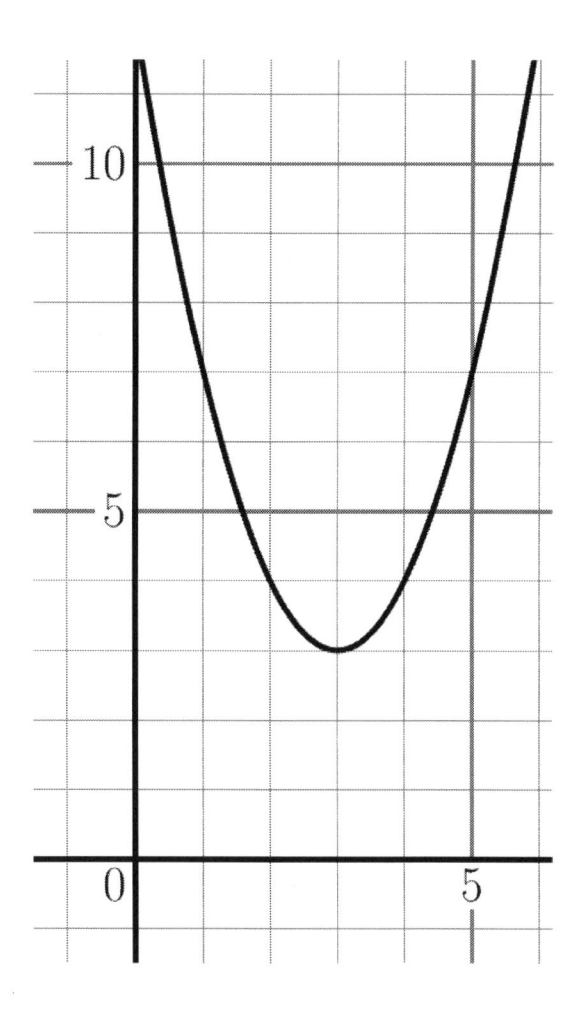

8. Which of the following could be the equation of the graph shown above?

A) $f(x) = x^2 - 6x + 12$
B) $f(x) = x^2 + 6x + 12$
C) $f(x) = x^2 - 6x - 12$
D) $f(x) = -x^2 - 6x + 12$

These choices use function notation where $y = f(x)$. We are shown an upward opening parabola with a vertex that appears to be at or very close to $(3, 3)$. In vertex form, the equation would be $(x - 3)^2 + 3$. The choices are not in this form, so we expand it by squaring $(x - 3)^2$. That gives us $x^2 - 6x + 9$ and then adding on that 3: $x^2 - 6x + 12$. This is choice A. No other choice is close to this one, so we can be comfortable choosing it.

Again, you could graph this one, and once you see that choice A—which can be entered into the OSC just as it is—looks just like the figure, you should be comfortable choosing it. Of course, if you have to try several choices, graphing each one, that could eat up time. **Know both ways to solve**, and then choose which one seems most efficient.

9. A researcher uses the equation $t = 3(m - 4)^2 + 21$ to model the temperature (t) in degrees of a solution m minutes after the beginning of an experiment. Which of the following is the best interpretation of the vertex of the graph of this equation?

Drill Explanations | 103

A) The maximum temperature of 4 degrees was reached 21 minutes after the beginning of the experiment.
B) The minimum temperature of 4 degrees was reached 21 minutes after the beginning of the experiment.
C) The maximum temperature of 21 degrees was reached 4 minutes after the beginning of the experiment.
D) The minimum temperature of 21 degrees was reached 4 minutes after the beginning of the experiment.

This equation is given in the vertex form: $a(x-h)^2 + k$. The vertex is therefore $(4, 21)$. In this scenario, the minutes are measured on the *x*-axis, and temperature is on the *y*-axis. This means that 4, as the *x*-coordinate of the vertex of an upward opening parabola, is the **minimum** value of the number of minutes. In other words, at 4 minutes, the temperature is 21. Before that time and after that time, the temperature is greater. Therefore, **choice D** is correct.

10. What is the *x*-intercept of the graph of $y = -5x - 6$?

To get the *x*-intercept of this line, we'll set *y* equal to 0. That is where the line crosses the *x*-axis. $0 = -5x - 6$. That simplifies to $5x = -6$ and $x = -\frac{6}{5}$ or -1.2. You could also graph this using the initial equation on the OSC and quite quickly see the location of the *x*-intercept.

11. Line *l* passes through points $(1, 4)$ and $(-2, 1)$. At what point(s), if any, does the graph of $f(x) = x^2 - 9$ intersect line *l*?

This situation presents us with a line and a parabola. We have all the information we need on the line, given those 2 points. You could set up 2 linear equations in order to determine the slope and *y*-intercept. $4 = (1)m + b$ and $1 = -2m + b$. The 1st equation tells us that $b = 4 - m$, and the 2nd equation tells us that $b = 2m + 1$. This means that $4 - m = 2m + 1$, and therefore $m = 1$. That lets us easily determine that $b = 3$. The equation of line *l* is therefore $y = x + 3$. (We don't need to write out that $m = 1$.)

Our 2 equations are now $y = x + 3$ for the line and $f(x) = x^2 - 9$, or simply $y = x^2 - 9$. To find where they intersect, yes, **we could graph them on the OSC and easily see where they intersect**.

Or, using our algebra, we could say that $x + 3 = x^2 - 9$ since each of these equals *y*. You might recognize the right side of that equation as the difference of 2 squares, which factors to $(x - 3)(x + 3)$. That would lead us to $x + 3 = (x - 3)(x + 3)$. But here you have to be careful. If you divide each side by $x + 3$, that seems to give us $1 = x - 3$, in which case $x = 4$, **but** this answer is not complete. There's another solution for *x*. The reason we got an incomplete solution is that we made an "illegal" move. We divided by $x + 3$ without knowing whether $x + 3$ could be 0. We **never** divide by 0.

So let's back it up to $x + 3 = x^2 - 9$. To solve this properly, we'll do what we typically do with quadratics: set it equal to 0 and factor. That gives us $x^2 - x - 12 = 0$. This factors to $(x - 4)(x + 3) = 0$, and **this** gives us both solutions for *x*: 4 and -3. We can plug these values into either of the original equations to find the *y*-coordinates. When *x* is 4, *y* is 7, and when *x* is -3, *y* is 0. The 2 intersecting points are therefore $(4, 7)$ and $(-3, 0)$.

Again, you want to have the skills to solve this algebraically, but in this case, **graphing on the OSC is a particularly efficient approach.**

12. Line l is perpendicular to a line that goes through points $(1,5)$ and $(3,0)$. What is the slope of line l?

The line that goes through these points has a slope of $-\frac{5}{2}$. How do we know that? You *could* graph it, but you could also just think, "the difference in the y values divided by the difference in the x values." That's $\frac{5-0}{1-3} = -\frac{5}{2}$. Of course, you'll avoid this trap choice or a trap choice that only shows the opposite of this value or the reciprocal of this value. We want the negative reciprocal, which is $\frac{2}{5}$.

13. Point B is the midpoint of points A and C. If the coordinates of B and C are $(7,-2)$ and $(11,4)$, respectively, what are the coordinates of A?

We know that the coordinates of a midpoint are simply the averages of the endpoint coordinates. But **be careful** because we are **given** the midpoint $(7,-2)$, as well as one of the endpoints $(11,4)$. We'll let (x, y) be the coordinates of A and set up 2 equations, 1 for each coordinate. That gives us $\frac{11+x}{2} = 7$ and $\frac{4+y}{2} = -2$. Solving the 1st equation gives us $x = 3$, and solving the 2nd one gives us $y = -8$. So the coordinates of point A are $(3,-8)$.

Day 9

Coordinate Geometry – Lines, Parabolas, Rational Functions, Circles

After reading this material and completing the drill at the end, sign and check the box for Day 9.

We've already worked on problems that combined algebra and geometry, and we'll now look at this sort of material in more depth. These problems are often found toward the end of a section where the more difficult questions are, or at least where they are **supposed** to be. We'll see that some of these supposedly harder questions aren't so bad if we approach them the right way. As we'll see, the "right way" often involves the use of the wonderful OSC available to you throughout both math modules.

Lines

By now, you certainly know the linear equation $y = mx + b$ where m is the slope and b is the y-intercept. For any given line, those values are **constants**; they don't change. On the other hand, x and y are variables; they refer to the coordinates of all the infinite points on a line. Let's say you're dealing with the equation $y = 3x - 1$. The slope of the line is 3, and it crosses the y-axis at -1. You can easily enter this equation into the OSC and see the graph of this line.

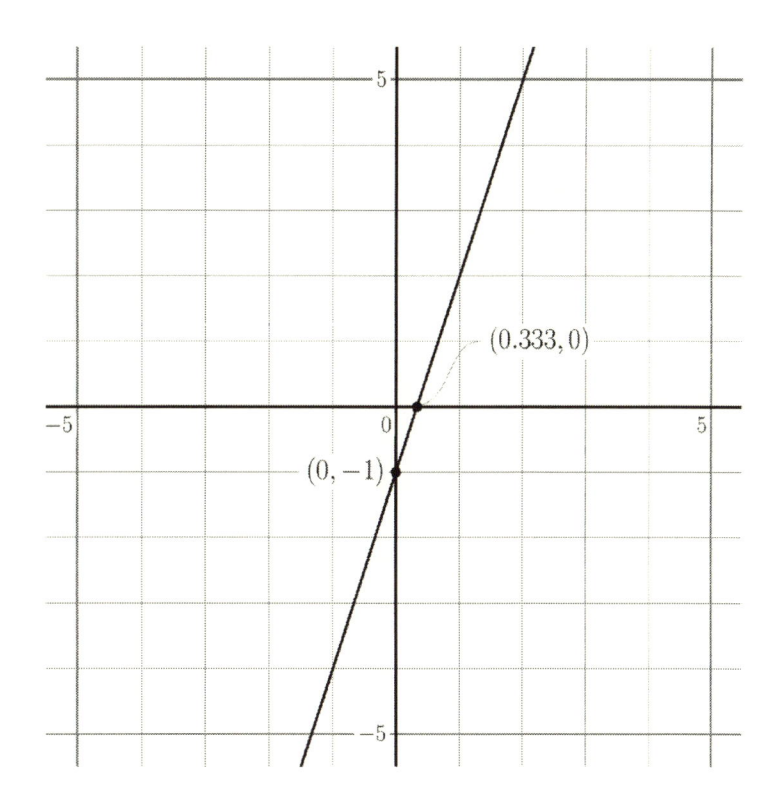

In a very real sense, the algebraic equation and the graph are the **same**. They present the same information. You will probably need to get the zoom level right, but once you do, you'll see both the *y and x* intercept. If they don't show up as gray points, click on the line, and they'll appear. Hover over them or click on them, and you'll see their exact locations. You can use their coordinates to determine the slope of the line, whether you think of it as "the difference of the *y* values over the difference of the *x* values" or as "rise over run." Note: the value .333 is a decimal approximation of 1/3. This is a common decimal/fraction equivalency that you should know. If you get a less common one such as 0.175, you can easily convert it to a fraction with the OSC. If you type the number into the space bar and then hit the little symbol that looks sort of like a fraction to the left of it, you'll see, in this case, $\frac{7}{40}$.

One advantage of the OSC is that you don't have to enter the equation of a line in the $y = mx + b$ format. For example, if you graph the equation $2(y + 1) = 6x$, you'll see that it's the same as the one just mentioned, $y = 3x - 1$. You can algebraically prove this, but the point is that you can just enter it in this more complicated form if that's how it looks on the test.

Let's now say that you need to solve for *y* given that $2(y + 1) = 6x$ and $2x - y = -5$. We know these are both lines, and without any exponents or roots, we have 2 linear equations. You should be able to solve them algebraically using substitution or elimination, but you could also just enter them **as is**, set the zoom level, scroll around the graph to see where the lines intersect, click there, and then see that the point where they intersect is $(6, 17)$. That means $y = 17$. With a little practice, this should take no more than about 15 seconds.

Coordinate Geometry – Lines, Parabolas, Rational Functions, Circles

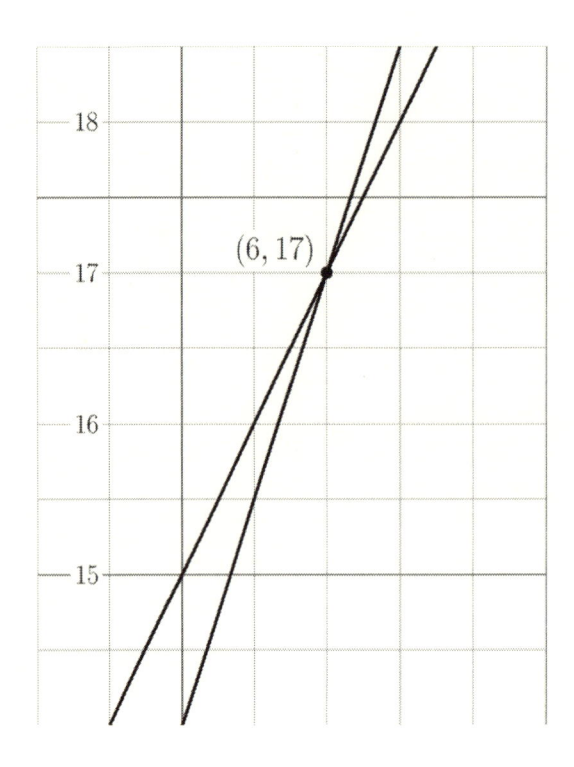

Parabolas

Parabolas correspond to quadratics, as we've seen. Typically, they will appear either in a form such as $x^2 + 8x + 12$ or in vertex form such as $(x + 4)^2 - 4$. (If you don't recognize that these equations are equivalent, go back to Day 4.) No matter how they appear, you can easily enter the quadratic into the OSC. The graph identifies the vertex and the location of any x and y intercepts. You can also determine these points from the equations, but the graph shows where they are at a glance.

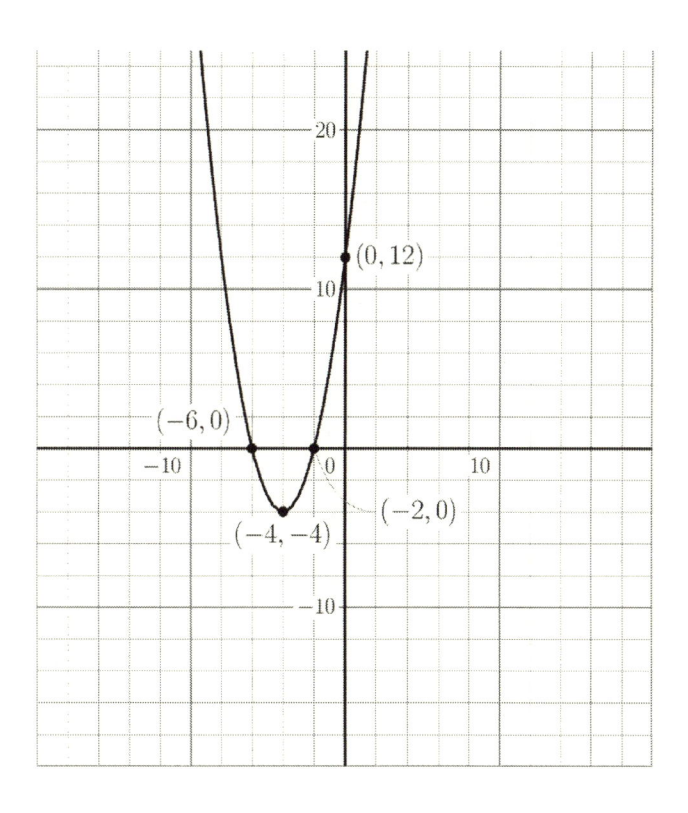

That's convenient, right?

Now let's say you have a question that asks for the intersection points of a line and a quadratic. We'll use the linear equation that we've used above, as well as the quadratic we just looked at, now set up to equal *y*. A question might say the following:

In the *xy*-plane, the graph of $y = 3x - 1$ intersects with the graph of $y = x^2 + 8x + 12$ at how many points, if any?

A) None
B) 1
C) 2
D) 3

You could use substitution to combine the equations into $3x - 1 = x^2 + 8x + 12$, and off you go, combining like terms, giving you $x^2 + 5x + 13 = 0$, but you'll soon see that there's no obvious way to factor and solve it. You might, therefore, try using the quadratic formula. But **before you do**, let's enter the graphs into the OSC. Actually, before doing even that, let's glance at the choices, which is always a good idea. A line can intersect with a parabola not at all, once, twice, and that's it. There's no way for them to intersect more than twice, so you can eliminate choice D. If you do graph the 2 equations, you'll see that they don't intersect at all. Choice A is correct.

If you **had** used the quadratic formula, you'd see that the *discriminant* (remember that from Day 4?) is negative, meaning that there is no real solution to the quadratic. No solution means no intersection.

Now let's look at a question that would be considered difficult on the exam.

Coordinate Geometry – Lines, Parabolas, Rational Functions, Circles | 109

In the xy-plane, a parabola with equation $y = -x^2 - 2x$ and a line with equation $\frac{y}{4} = k$ for some constant k intersect at exactly 1 point. What is the value of k?

We'll say that this is **not** a multiple choice question. You have to enter the solution. As always, we can solve this algebraically, but let's try the graphic approach first. If we enter the quadratic equation, we get the following graph.

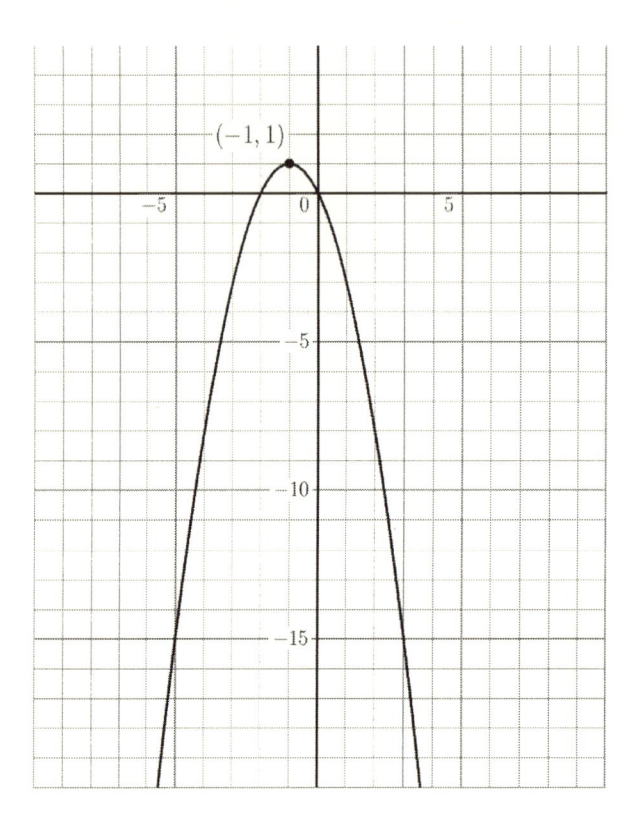

By clicking on the vertex, we can see its coordinates, $(-1, 1)$. This is important for this equation because if a line with a slope of 0 intersects a parabola, they intersect at the parabola's vertex. How do we know the line has a 0 slope? The equation is $\frac{y}{4} = k$. We can simplify that to $y = 4k$. Whatever k is, that means the line is parallel to the x-axis and therefore has a slope of 0. **That** means it must intersect the parabola at its vertex. Since we can see that the vertex has a y-coordinate of 1, we can say that $1 = 4k$, which means $k = \frac{1}{4}$. You can enter the answer this way, or as .25.

Before looking at some more "graphables," let's remember that not all problems that *could* be solved by graphing *should* be solved by graphing. If, for example, you need the vertex of the parabola with the equation $y = (x + 2)^2 + 10$, there's no point in graphing. This is the vertex form, so it's obvious that the vertex is at $(-2, 10)$. Let's look at something not so obvious.

$$f(x) = 3x^2 + bx - c$$

In this equation, b and c are constants. The graph of $y = f(x)$ is a parabola with a vertex at the point (h, k). If $f(2) = f(-6)$, what is the value of b?

110 |Day 9

There's no obvious way to graph the parabola because we don't know the value of b or c. But we do know a thing or 2 about the parabola. It opens upward since 3 is positive. Also, since $f(2) = f(-6)$, we know that the points with x-coordinates 2 and -6 have the same y-coordinate. **That means they are equidistant from the vertex**. The average of 2 and -6 is negative 2, so h, which is the x-coordinate of the vertex, is -2.

The next step is to apply the equation that gives us the x-coordinate of the vertex, $-\frac{b}{2a}$. We **know** the x-coordinate of the vertex, and we **know** that a is 3, so we can say $-\frac{b}{2(3)} = -2$. This simplifies to $12 = b$. We're done.

Rational Functions – Functions with Fractions

Sometimes we see rational functions where polynomials are found on the top and bottom of a fraction. On the test, these will usually be fractions with an unknown on the bottom. They are typically graphed as curves. Here's a high-level question that does include a graph but still requires some algebra.

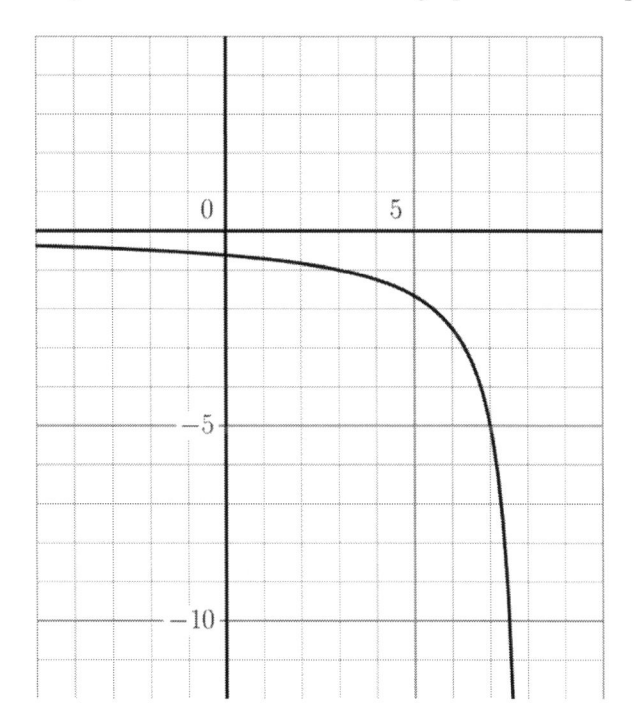

The rational function f is defined by the equation $f(x) = \frac{n}{x-p}$ where n and p are constants. The partial graph of $y = f(x)$ is shown. If $g(x) = f(x-3)$, which equation could define function g?

A) $g(x) = \frac{x}{16}$
B) $g(x) = 11$
C) $g(x) = \frac{5}{x-11}$
D) $g(x) = \frac{16}{x-11}$

They've given us a graph, but there's no obvious way to do our own graphing. We can, however, get some useful information from the graph that's shown. We can see that $(3, -1)$ and $(7, -5)$ appear to be on the graph. Of course we can't know this for sure, but these points are at least very close to being on the graph, and the fact that the values are all integers makes them likely.

We can now plug these values into the function f. That gives us $-1 = \frac{n}{3-p}$ and also $-5 = \frac{n}{7-p}$. We're now in good shape since we have 2 equations with 2 unknowns. We can **carefully** cross-multiply each of them to get $n = p - 3$ and $n = 5p - 35$. That means $p - 3 = 5p - 35$, which means $4p = 32$ and $p = 8$.

Now let's go back and look at what we wanted, which was an equation for $g(x)$. Since $g(x) = f(x - 3)$, we can say that $g(x) = \frac{n}{x-3-p}$. (We just took our original equation for f and replaced x with $x - 3$.) Since we've solved for p, we can simplify this to $g(x) = \frac{n}{x-3-8}$ or simply $g(x) = \frac{n}{x-11}$. Choices C and D are starting to look real good. In fact, this question would certainly be considered hard enough to be at the end of a Math section, and you might need to do a little intelligent guessing at this point. **NEVER, NEVER FORGET** to "peek" at the choices occasionally for "hints" to the right answer, especially when you're short on time.

But let's wrap this up. It won't take long. The only difference between choices C and D is the numerator. The numerator of $g(x)$ is n, and we can easily determine n using one of the equations we've created (and the fact that we know that $p = 8$). If we use the equation $-1 = \frac{n}{3-p}$ and substitute 8 for p, we get $n = 5$. **Choice C** is correct.

Circles: $(x - h)^2 + (y - k)^2 = r^2$

You should have memorized this important equation, but let's look at a case where the OSC can once again give us a helping hand.

What is the length of the circumference of the graph of $x^2 + 4x + y^2 - 6y = 51$ in the xy-plane?

To solve this algebraically, we'll need to get that equation into the standard form given above. That involves "completing the square," which we know how to do. Right? But the OSC doesn't care about form. We can simply enter the equation as given in the question, which will give us the following circle.

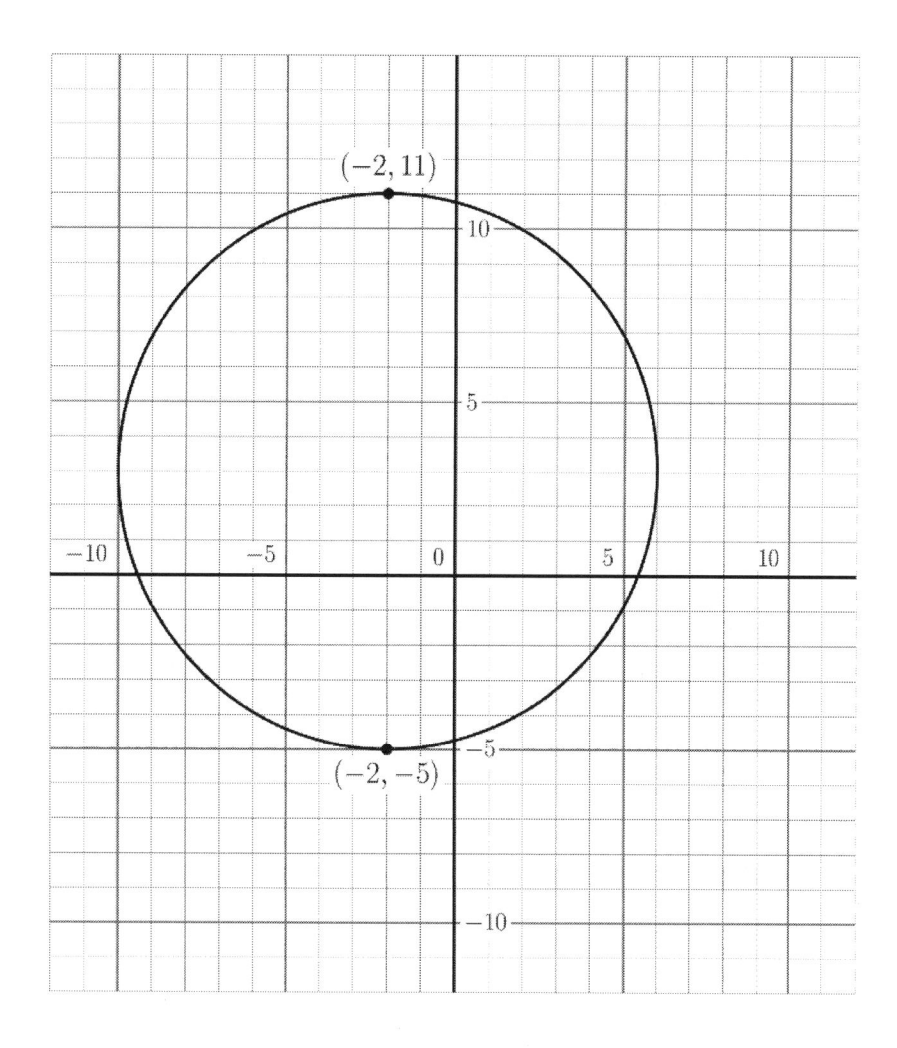

Using the labeled points, we can easily see that the circle has a diameter of 16. That means the circumference is 16π. We're done.

Day 9

Drill

Complete all the following questions.

1. What is the x-intercept, y-intercept, and slope of the line with the equation $2.5x + \frac{y}{2} = 3$?
Note: Find the answers using algebra **and** the OSC.

2.
$$1.5y = 4.5x - 6$$
$$x + y = -10$$

Given the system of equations above, what is the value of $x - y$?
Note: Find the answer whichever way you prefer.

3. What is the x-coordinate of the vertex of a parabola with the equation $y = x^2 - 14x + 47$?

4. What is the y-intercept of a parabola with the equation $y = 3(x + 4)^2 - 6$?

5. The point $(3, 21)$ lies on a parabola with the equation $y = 2x^2 - 4x + 15c$ where c is a constant. What is the value of c?

6. The functions $f(x) = -(x + 2)^2 + 8$ and $g(x) = 2x^2 - 19x - 20$ are graphed on the xy-plane. What is the sum of the x-coordinates of $f(x)$ where it equals $g(x)$? (Round the answer to the nearest integer.)

7. The point $(2, 1)$ lies on line l, which has a y-intercept of 4. Line m is perpendicular to line m. What is the slope of line m?

8. Line w in the xy-plane has an x-intercept at 2. The line intersects a parabola with the equation $y = -(x + 2)^2 + 10$ at $(-4, 6)$ and at (r, s). What is the sum of r and s?

9.
$$f(x) = 4x^2 + bx + c$$

The given equation defines the function f where b and c are constants. The graph of the equation has a vertex at $(3, -1)$ and x-intercepts at $(2.5, 0)$ and $(a, 0)$. What is the value of a?

10.

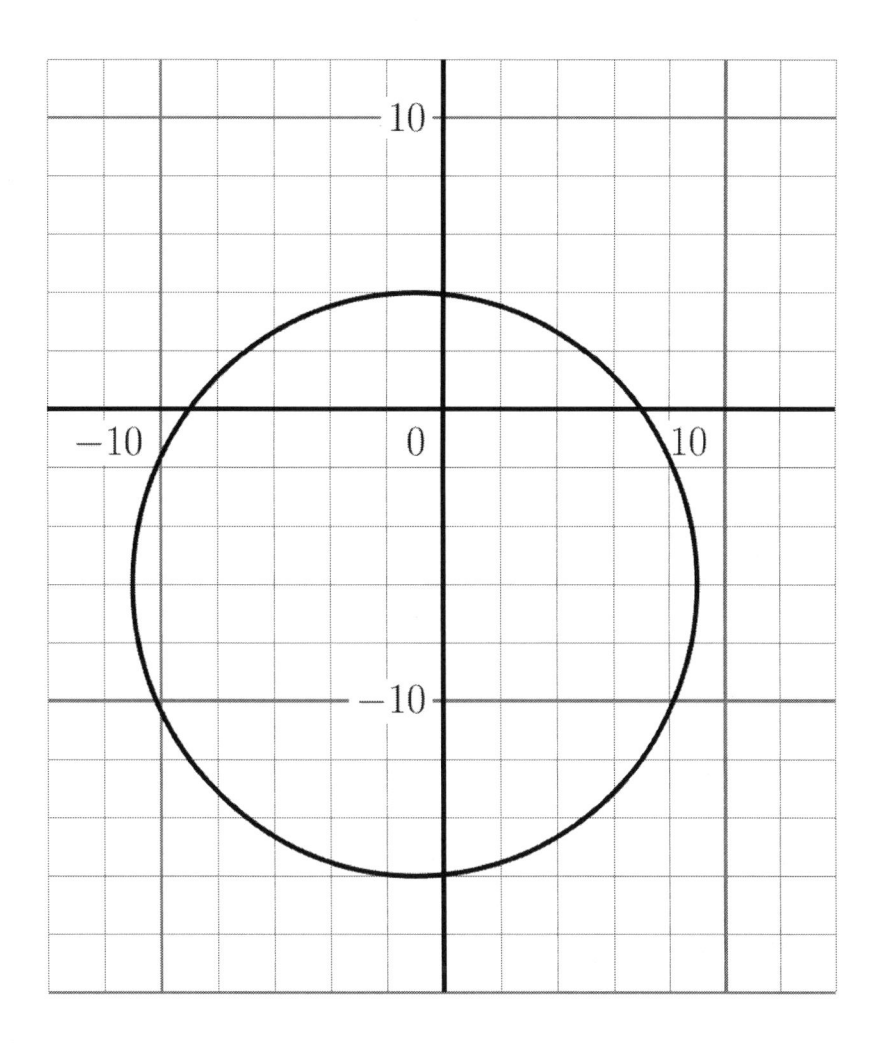

The circle above is represented by the equation $(x + 1)^2 = c^2 - (y + 6)^2$ for some constant c. Which of the following could be equal to c?

A) 5
B) 10
C) 10π
D) 100

Drill | 115

11.

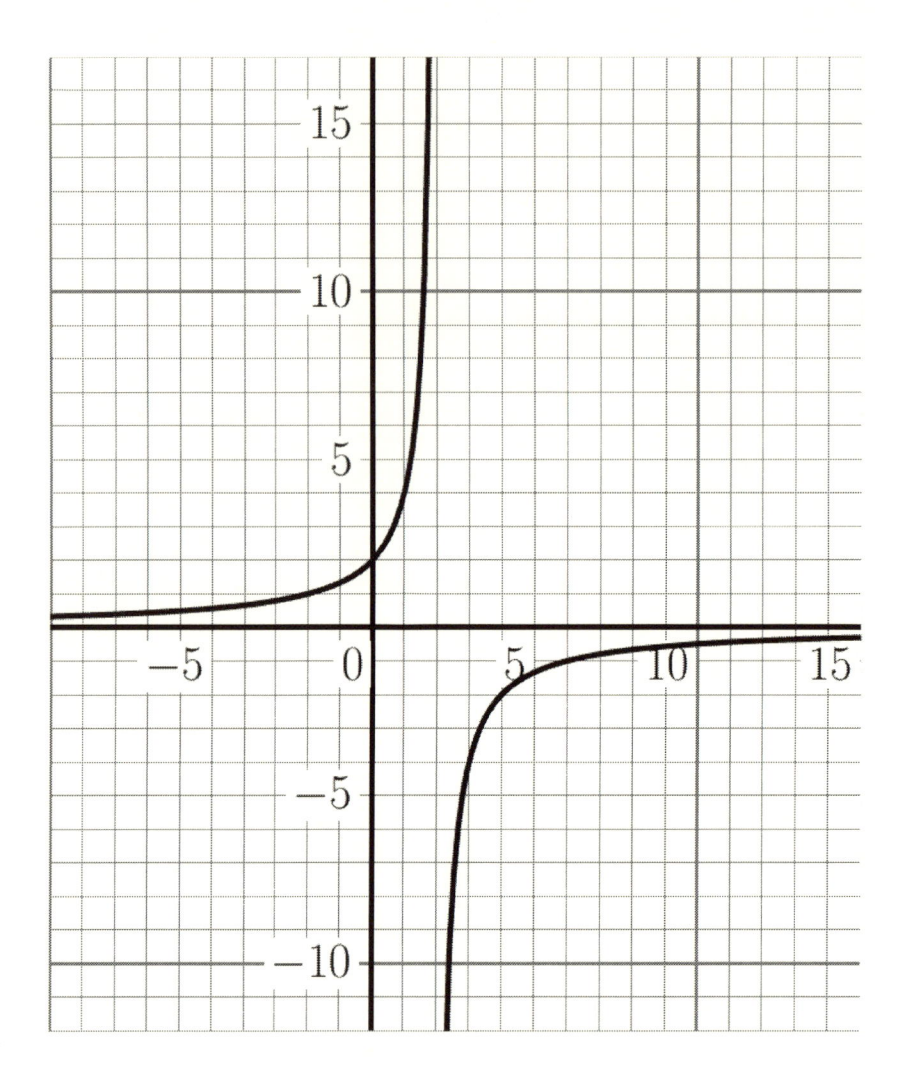

Which of the following could be the equation of the graph shown in the xy-plane?

A) $\dfrac{1}{x-2}$
B) $\dfrac{x+2}{4}$
C) $\dfrac{4}{x-2}$
D) $\dfrac{4}{x} - 2$

Day 9

Drill Explanations

1. What is the x-intercept, y-intercept, and slope of the line with the equation $2.5x + \frac{y}{2} = 3$?
Note: Find the answers using algebra **and** with the OSC.

Algebraically, you can determine the x-intercept by setting y to 0. Likewise, you can determine the y-intercept by setting x to 0. That tells us the x-intercept of this line is 1.2 and the y-intercept is 6. Knowing that the coordinates (1.2, 0) and (0, 6) are on the line allows us to determine that the slope is −5.

But we could also use the OSC to efficiently graph these 2 lines. Once we do, we can easily see where the intercepts are and then, as before, use these to determine the slope.

2.
$$1.5y = 4.5x - 6$$
$$x + y = -10$$

Given the system of equations above, what is the value of $x - y$?

Algebraically, you might decide to simplify the top equation by doubling every term, giving you $3y = 9x - 12$. You could further simplify by dividing each of these terms by 3, giving you $y = 3x - 4$. You might now use substitution, plugging in $3x - 4$ into the other equation. From there, you can solve for x and y.

Using the OSC, you could graph these 2 lines. The intersection occurs at $(-1.5, -8.5)$. Since we want $x - y$, that's $-1.5 - (-8.5) = 7$.

3. What is the x-coordinate of the vertex of a parabola with the equation $y = x^2 - 14x + 47$?

To get the answer algebraically, all that's needed is the vertex equation, $-\frac{b}{2a}$. That gives us $-\frac{-14}{2} = 7$. You could also graph it and read the vertex off the OSC.

4. What is the y-intercept of a parabola with the equation $y = 3(x + 4)^2 - 6$?

Algebraically, you could simply set x equal to 0, giving you $y = 42$. You could also graph it on the OSC. Once you click on the graph, the intercepts and vertex will appear.

5. The point (3, 21) lies on a parabola with the equation $y = 2x^2 - 4x + 15c$, where c is a constant. What is the value of c?

This one is best handled algebraically since we can't graph the parabola without knowing c. If you plug in the coordinates of the given point, you get $21 = 2(3)^2 - 4(3) + 15c$, which simplifies to $c = 1$.

6. The functions $f(x) = -(x+2)^2 + 8$ and $g(x) = 2x^2 - 19x - 20$ are graphed on the xy-plane. What is the sum of the x-coordinates of $f(x)$ where it equals $g(x)$? (Round the answer to the nearest integer.)

This one is best handled graphically. If you graph the two parabolas, you'll see that the x-coordinates at the two points of intersection are 6.275 and -1.275. The sum of these is 5. You could set the 2 equations equal to each other, although you'll get a quadratic that is not easily solved. (If you did go down this road, you'd see that the values derived from our algebraic solution are very close approximations, but remember that the answer only asks for the nearest integer, so we're safe.)

7. The point $(2, 1)$ lies on line l, which has a y-intercept of 4. Line m is perpendicular to line l. What is the slope of line m?

We can do this efficiently by plugging the given values into the linear equation $y = mx + b$. That gives us $1 = m(2) + 4$, which simplifies to $m = -\frac{3}{2}$. Since m is perpendicular to this line, its slope is $\frac{2}{3}$.

8. Line w in the xy-plane has an x-intercept at 2. The line intersects a parabola with the equation $y = -(x+2)^2 + 10$ at $(-4, 6)$ and at (r, s). What is the sum of r and s?

Combining your algebra skills with the OSC can take care of this problem. The parabola can be easily entered and graphed on the OSC. You can also set up 2 equations for the line. We have the given value of $(-4, 6)$, which gives us $6 = m(-4) + b$, and, since the x-intercept is 2, we also have $0 = m(2) + b$. Combining these 2 equations lets us solve for m and b. We get $m = -1$ and $b = 2$. You can now graph the line $y = -x + 2$. Once we do this, we see that the graphs intersect at $(1, 1)$. That gives us a sum of 2.

9.
$$f(x) = 4x^2 + bx + c$$

The given equation defines the function f where b and c are constants. The graph of the equation has a vertex at $(3, -1)$ and x-intercepts at $(2.5, 0)$ and $(a, 0)$. What is the value of a?

The vertex of a parabola is midway between the 2 x-intercepts. Since the x-coordinate of the vertex is 3 and 1 x-intercept is at 2.5, the other x-intercept is at 3.5.

10.

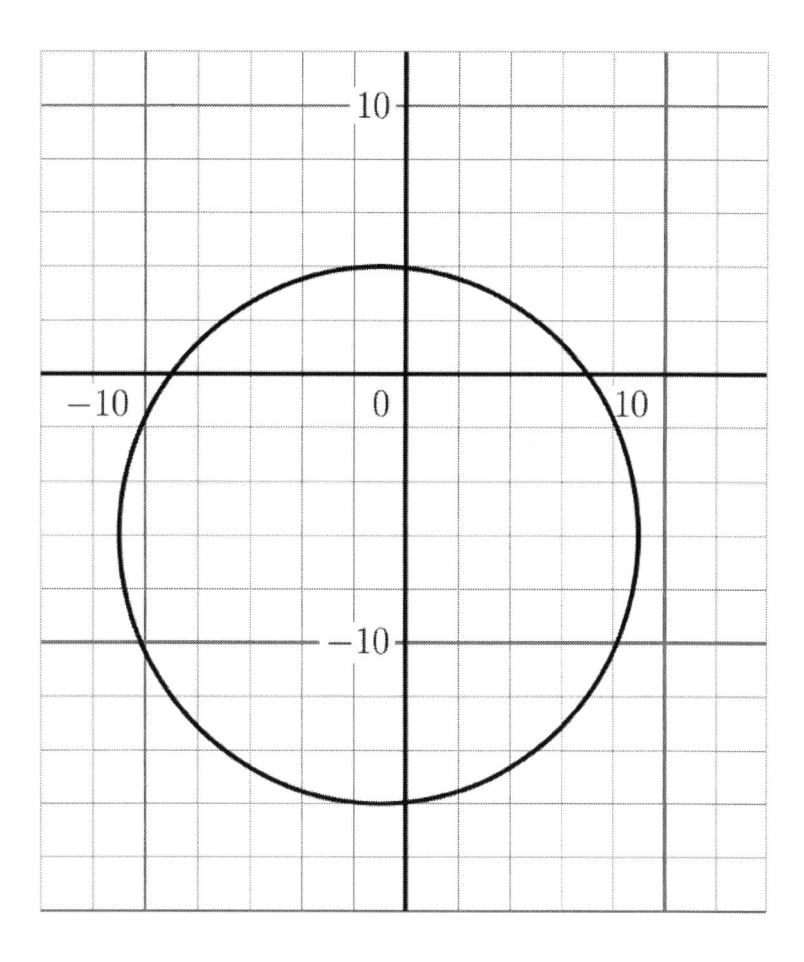

The circle above is represented by the equation $(x + 1)^2 = c^2 - (y + 6)^2$ for some constant c. Which of the following could be equal to c?

A) 5
B) 10
C) 10π
D) 100

We can easily get this equation into the standard form of the equation of a circle by adding $(y + 6)^2$ to each side. That gives us $(x + 1)^2 - (y + 6)^2 = c^2$. In this form, the radius is c. If you look at the choice and the graph, only 10 could be the radius.

11.

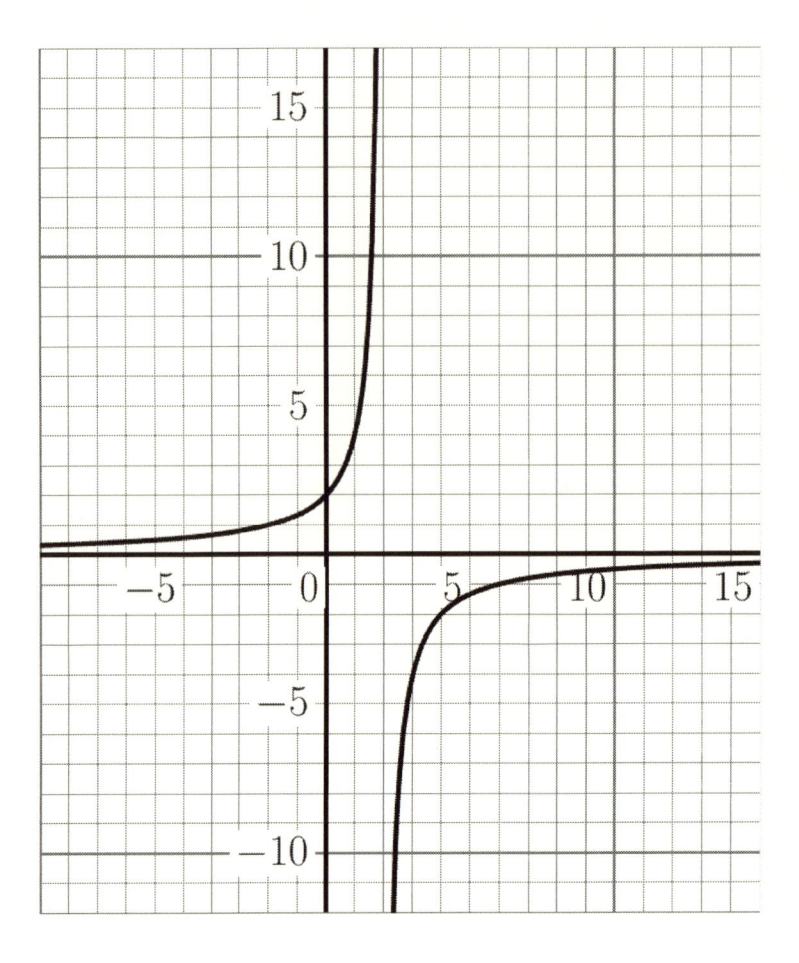

Which of the following could be the rational equation of the graph shown in the *xy*-plane?

A) $y = \frac{1}{x-2}$
B) $y = \frac{x+2}{4}$
C) $y = \frac{4}{2-x}$
D) $y = \frac{4}{x} - 2$

One way to approach this problem is to realize that both parts of the graph appear to approach an *x* value of 2 without ever reaching it. (This is called an *asymptote*.) Why can't it reach 2? Well we can never divide by 0, and choices A and C show us a fraction that **would** divide by 0 if *x* were equal to 2. One of those has to be correct.

We can now try some values that appear to lie on the graph. For example, $(0, 2)$ appears to lie on the graph. Would those values work in any of the choices? Yes, just the correct one: **C**.

Day 10

Word Problems 2 – Representation, Compound Interest

After reading this material and completing the drill at the end, sign and check the box for Day 10.

Today, we'll look at the kind of real-world word problems that show up on the test and do **not** require you to solve for an unknown. Instead, you'll need to understand the information **represented** by an algebraic term or equation. In these situations, the OSC is typically less helpful.

Let's start with a line. A line might not seem to convey much information, but it does. It has a crucial slope and 2 intercepts. (All right, it's true that horizontal and vertical lines only have 1 intercept.) In particular, *slope* represents a *rate of change*, which has a place in the real world. For example, a new business that sells notepads might have more and more customers during its first 6 months. This is an increasing rate of change. The more time that "increases," the more customers it has. (Of course, it's unlikely that the rate increases in a perfectly linear manner, but you get the idea.) Or let's say that during a scientific experiment, the number of bacteria in a dish decreases over time. That is a decreasing rate of change. The graph of that could like this:

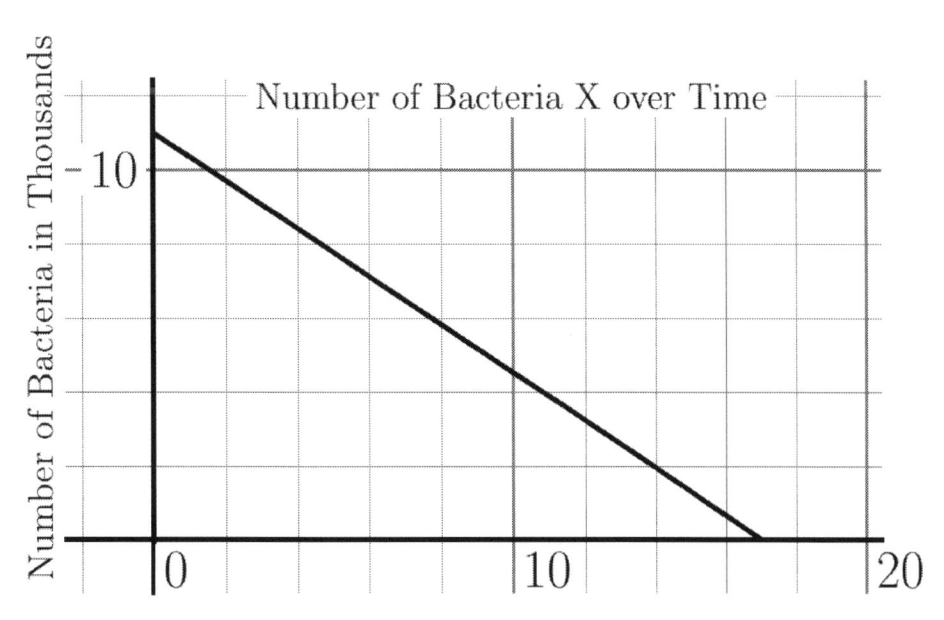

A few things are worth pointing out. First, this is not actually a line; it's a line segment. At the start of the experiment, when 0 hours have gone by, there are somewhat more than 10,000 bacteria. (Of course, we must note the "in thousands" label on the y-axis.) This is the *y*-intercept. Between 16 and 18 hours, there are 0 bacteria. This is seen at the *x*-intercept. We clearly have a decreasing rate of change, which is to say a negative slope. Without any other specific values, we couldn't calculate the slope, although we can approximate it. The segment decreases about 11,000 over a change of about 17 hours. That gives us a slope of about $-\frac{11,000}{17}x$, which is close to $-650x$.

We can also approximate the *y*-intercept at 11 (thousand), giving us the equation $y = -650x + 11,000$. Since this is a segment (not a line), we should also say $0 \leq x \leq 18$.

Now let's see how the SAT might present a question of this sort.

Researchers investigated the number of bacteria in thousands (*n*) remaining in a Petri dish *h* hours after the start of the experiment. They created the graph above to show the results of the experiment. Which of the following could be the equation representing this graph?

A) $n = 650h + \frac{53}{5}$
B) $n = -650h + 11,000$
C) $n = -\frac{53}{5}h + \frac{3}{5}$
D) $h = -650n + 11,000$

We now need to think of *n* as the y-value and *h* as the x-value. Only 1 of the choices comes close to our approximation. But let's say you didn't bother to set up this approximation. You can still efficiently narrow down the choices to the right one. Choice D has the *h* and *n* reversed. Choice A shows a positive slope. Choice C shows values that don't take into account the "thousands" of the *y* values. That leaves **choice B**.

Also, it's not a bad idea to have some idea of what slopes look like. For example, a positively sloped line that cuts through the origin (0, 0) and bisects the quadrants has a slope of 1. If the line is steeper than that, its slope is greater than 1, and vice versa. A negatively sloped line that cuts through the origins and bisects the other 2 quadrants has a slope of negative 1. Note: These slope approximations are only valid if the scales on the 2 axes are the same. If an axis increases by 1 with each tick mark and the other increases by 10 with each tick mark, this won't work.

Let's jump to a word problem that involves *planar* geometry as opposed to the *coordinate* geometry we've been looking at recently. Again, we're not going to solve; we're going to *represent*.

Angie lives *x* miles due west of her friend Enzo and *y* miles due south of her friend Bea. Which of the following expressions represents the distance between Bea's house and Enzo's house?

A) $\sqrt{x^2 + y^2}$
B) $x^2 + y^2$
C) $x^2 - y^2$
D) $\sqrt{x^2 - y^2}$

This question should make you happy because each house is 1 vertex of a right triangle, and we know **a lot** about right triangles. You could sketch this out in a few seconds, and there you go—a right triangle with x and y as the legs and with the other side the hypotenuse. We can't use our knowledge of the special right triangles in this case, so we'll have to depend on the Pythagorean theorem, which tells us that $a^2 + b^2 = c^2$. If we plug in the values of x and y for a and b, get c by itself, and **then** take the square root of each side, we'll get choice A.

Let's now try a rate mixed with an algebra question where we once again need to represent.

On his way to work, Luis drove at an average speed of 30 miles per hour for r miles and then at an average speed of 50 miles per hour for s miles until he got to work. If the total number of hours he spent driving to work is t, which of the following expressions represents this situation?

A) $\frac{r}{30} + \frac{s}{50} = t$
B) $30r + 50s = t$
C) $\frac{30}{r} + \frac{50}{s} = t$
D) $\frac{t}{30} + \frac{t}{50} = r + s$

Let's remember the rate formula: $rt = d$, meaning rate x time = distance. Of course, you should be comfortable expressing this as time = distance/rate **or** as rate = distance/time. In this case, looking at the choices, we see that 3 of them have expressions that equal t, the total time. So, if 1 of those is correct, the other side of the equation should represent the **sum** of the time of each part of the trip to work. Since we want *time*, we want distance over rate. **Choice A** gives us exactly that. Notice that we're not solving anything (we can't with 2 unknowns and only 1 equation) but just setting up an equation. (It's a tricky one since r is used for *distance*.)

This can also happen with another important 3-part equation: Average = Sum/Number of Terms, which can also be expressed as Sum = Average x Number of Terms or as Terms = Sum/Average.

Painters working on a building estimate that they'll use v cans of paint for the first 6 days of work and then w cans of paint for the remaining 5 days of work. Using this estimate, they conclude that they'll need 124 cans of paint. Which of the following expressions represents this situation?

A) $\frac{6}{v} + \frac{5}{w} = 124$
B) $\frac{11}{v+w} = 124$
C) $6v + 5w = 124$
D) $30vw = 124$

We're given the sum of all the cans, so we'll need to represent the total number of cans used for the first 6 days and then the remaining 5 days. The equation we can best use is Sum = Average x Number of Terms. For the first 6 days, that gives us $6v$. For the other 5 days, the sum is $5w$. Adding those up, we get choice C.

Word Problems 2 – Representation, Compound Interest | 123

Compound Interest

This is a concept that we see in real life when we have money in a typical interest-bearing savings account. Let's say that a bank offers 4% interest annually. If you deposit $100 and don't withdraw or deposit any funds, at the end of the year you'll have an extra 4% of $100. In other words, you'll have an extra $4, giving you a total of $104. If again you don't add or subtract anything, after another year you'll have an extra 4% of $104. You'll now have $108.16. As we've seen before, when talking about percentages, the most direct way to calculate this would be to multiply 104 by 1.04.

So the 1st year, we multiply our initial deposit by 1.04. The 2nd year we multiply this new, bigger amount by 1.04 again. In other words, we multiply $100 by 1.04^2. After 3 years, we can just multiply $100 by 1.04^3. The exponent is simply the number of years. If instead the interest *compounded* twice a year, we would multiply our original amount by twice the number of years.

In the year 2024, the average cost of a house in City X is estimated to increase by 3.2% annually over the next 10 years. If the current average cost of a house in City X is h, which of the following represents the estimated average cost of a house in City X in 2029?

A) $5(1.032)^h$
B) $h(1.032)^5$
C) $h(0.32)^5$
D) $(5h)^{1.032}$

The initial cost is h. After 1 year, we would multiply this by 1.032 to get the new cost. After 2 years, we would multiply this new, bigger number by 1.032. More easily, we could just multiply the original cost, which is h, by 1.032^2. In this case, 2029 is 5 years later than 2024, so our exponent needs to be 5. We raise 1.032 by this power and then multiply it by the original cost of h. That's **choice B**.

Here's another word problem where our job is not to solve but to interpret.

An art museum is holding a craft-building event. Based on past attendance to similar events, the project leader estimates 23 people will attend. The function $f(x) = 2x + 50$ approximates the volume of glue in liquid ounces needed, given that the actual number of people who attend the event is $23 + x$. Which statement is the best interpretation of the y-intercept of the graph of $y = f(x)$ in the xy-plane in this context?

A) The museum will need 50 ounces of glue if 23 people attend the event.
B) The museum will need 2 ounces of glue if 50 people attend the event.
C) The museum will need 2 ounces of glue for every 50 people who attend the event.
D) The museum will need 25 ounces of glue if 23 people attend the event.

The y-intercept is where the x-coordinate is 0. If we plug 0 in for x, we get $f(0) = 2(0) + 50 = 50$. That means if $23 + 0$ people attend—in other words, exactly 23 people—50 ounces of the glue are needed. That's **choice A**.

Here's another interpretation question.

A tire store sells 3 kinds of tires, which they refer to as A, B, and C. During a weekend, they sold 21 A tires, 8 B tires, and 16 C tires for a total of $27,000. The equation $21A + 8B + 16C = 27,000$ represents this situation. Which is the best interpretation of 16C in this context?

A) The selling price of 1 C tire
B) The total selling price of all the C tires bought that weekend
C) The number of tires bought that weekend
D) The total selling price of the C tires bought each day of that weekend

We know that the right-hand side of the equation is the total selling price of all 3 kinds of tires. We also know that 16 of the C tires were bought. This means that C is the selling price of 1 C tire, and 16C is the selling price of all 16 of them. Therefore, **choice B** is correct.

Now we'll look at a problem that would be almost certainly too time-consuming to completely work out algebraically. That means we'll consider other (or additional) methods.

Biologists studying a microorganism that lives deep under water estimate that its lifespan in days when kept at an average temperature of t can be estimated as $m(t) = -6t^2 - 14t + 1$. Which of the following equivalent forms of $m(t)$ shows, as constants or coefficients, the maximum possible lifespan and the temperature that result in the maximum lifespan?

A) $6\left(t + \frac{7}{6}\right)^2 + \frac{55}{6}$
B) $-6\left(t + \frac{7}{6}t\right)^2 + \frac{55}{6}$
C) $-6t\left(t + \frac{7}{6}\right)^2 + \frac{55}{6}$
D) $-6\left(t + \frac{7}{6}\right)^2 + \frac{55}{6}$

As mentioned, converting the equation as given into its vertex form is time-consuming. It's best to realize that you **do want the vertex form**, and then scan the choices. Why the vertex form? That's the form that gives us the coordinates of the vertex, $a(x - h)^2 + k$, in which (h, k) are the coordinates of the vertex. The given equation represents a downward-opening parabola, so the vertex will give us what we want—the maximum possible lifespan (the k value) and the temperature that results in the maximum lifespan (the h value). Choice B is quickly ruled out because it doesn't give us a constant h. Instead, it gives us $-\frac{7}{6}t$, which varies depending on the value of t. Choice A is obviously wrong because it describes an upward opening parabola. And choice C is gone because it needlessly puts an extra t outside the parentheses. So without having to do the calculations, we can choose **D**.

Can graphing help us with this one? Yes. But first, remember that in using the OSC, you'll have to replace t with x. Once you do, you can enter the original equation and see that the vertex is at the decimal equivalent of $\left(-\frac{7}{6}, \frac{55}{6}\right)$. Just looking at the choices, we already know those are the relevant numbers. Continuing this way, however, you could now try each choice until it appears to be identical to the original equation's graph. This could take a while. Just recognizing the form, as explained above, would take less time.

Word Problems 2 – Representation, Compound Interest

Large, metal, right-circular cylinders are used to store liquids used by a factory. Each cylinder has a circumference of 2π meters and a height of 4 meters. If $f(d) = \dfrac{\pi r^2 h d}{60c}$ represents the number of minutes required to pour the liquid into d cylinders, what does the constant c represent?

A) The rate at which 1 cylinder can be filled in cubic centimeters per minute
B) The rate at which 1 cylinder can be filled in cubic centimeters per second
C) The fraction of 1 cylinder that can be filled in 1 minute
D) The time it takes to fill 1 cylinder in hours

For this question, we need to know our volume formula for a right circular cylinder, or $V = \pi r^2 h$. Since we're dealing with volume, that gives us **cubic** units (meters, miles, quarts, etc.). If we're filling d cylinders, then the volume needed is $\pi r^2 h d$. To get the **time** needed to pour that amount, we'll have to divide it by the **rate** at which it's poured. That tells us that the denominator must be a rate, making choices A and B look good. Since $60c$ is our rate **per minute**—remember, we're told that the equation give us *minutes*—then c must be the rate **per second**, which is choice B. That can be confusing. If you know how many seconds go by, you must **divide** by 60 to get the number of minutes.

Day 10

Drill

Complete all the questions below.

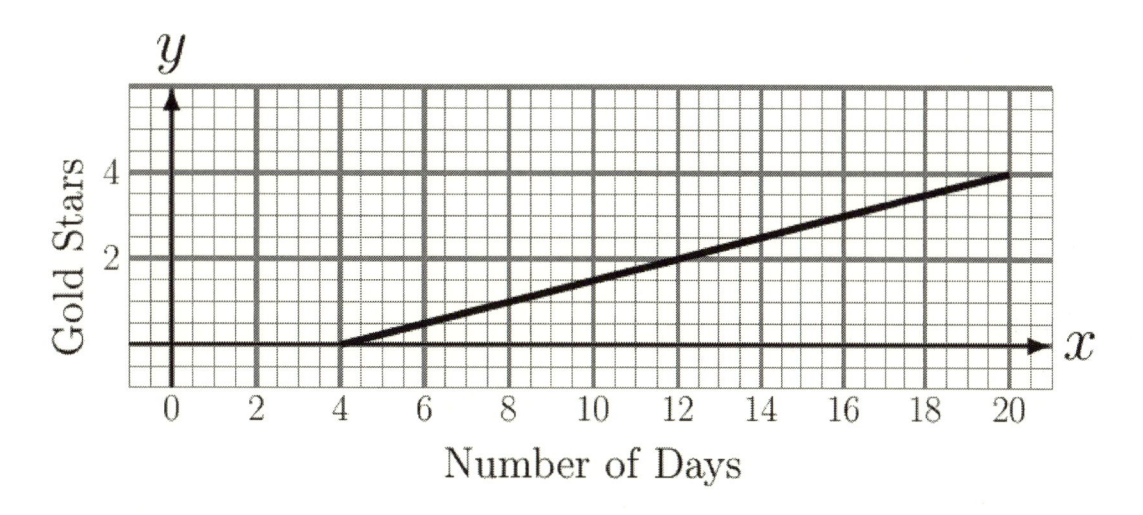

1. A 3rd grade teacher has devised a system for rewarding her students. Students who get all their work done for at least 8 consecutive days are awarded gold stars. The teacher then created the graph above to show the relationship between the number of days and the number of gold stars. Which of the following could be the equation representing this graph?

A) $y = \frac{4}{x} + 1$
B) $y = \frac{x}{4} - 1$
C) $y = \frac{x}{4} + 1$
D) $y = 4x - 1$

Note: The next 3 questions all relate to the situation in Question 2.

2. An economic analyst has been tracking the performance of the price of 1 share of a company's stock. They use the equation $f(x) = (x - 3)^2 + 9.50$ to approximate the price of 1 share x weeks from the coming Monday. What is the best interpretation of the y-intercept of the graph of $y = f(x)$ in the xy-plane in this context?

A) The price of 1 share of the stock this coming Monday will be $9.50.
B) In 3 weeks, the price of 1 share of the stock will be $9.50.

Drill | 127

C) In 3 weeks, the price of 1 share of the stock will be $18.50.
D) The price of 1 share of the stock this coming Monday will be $18.50.

3. What is the best interpretation of the coordinates of the vertex in this same context?

A)   In 3 weeks from the coming Monday, 1 share of the stock will reach its minimum price of $9.50.
B)   In 3 weeks from the coming Monday, 1 share of the stock will reach its minimum price of $18.50.
C)   In 3 weeks from the coming Monday, 1 share of the stock will reach its maximum price of $9.50.
D)   In 3 weeks from the coming Monday, 1 share of the stock will reach its maximum price of $18.50.

4. Why is there no x-intercept in this same context?

A) The equation does not show the price of 1 share of the stock on this coming Monday.
B) The equation does not show the current price of 1 share of the stock.
C) The price of 1 share of the stock will never reach a maximum value.
D) The price of 1 share of the stock will not reach 0.

5. Phil typically drives the 250 miles to his cousin's home at an average rate of r mph. Because of traffic, his average time making the trip most recently was $r + 6$ mph. Which of the following expressions represents the additional time it took Phil to make the trip?

A) $\frac{250}{r} - \frac{250}{r+6}$

B) $\frac{250}{r+6} - \frac{250}{r}$

C) $\frac{r}{250} - \frac{r+6}{250}$

D) $250r - 250(r - 6)$

6. Over 2 days, shoppers at a mall were asked to take part in a survey where they could rate how they felt about a soft drink on a scale of 1 to 12. On the 1st day, n people gave the drink an average score of s. The following day, $n - 10$ people gave the drink an average score of $s - 1$. If the sum of all the scores is 430, which of the following equations represents this situation?

A) $12ns + 10(n - 10) = 430$
B) $n(s - 1) + (n - 10)s = 430$
C) $ns + (n - 10)(s - 1) = 430$
D) $10ns + 10(ns - 1) = 430$

7. Four years ago, Saheed deposited x dollars into a savings account that generates compound interest biannually (twice a year). The expression $f(x) = (x)(1.0425^{2t})$ represents the amount in the account after a period during which no money was withdrawn or deposited. Which of the following is the best interpretation of t in this expression?

A) The amount that Saheed first deposited into the account
B) The number of years during which the account generates interest
C) The amount of money in the account after 2 years
D) The amount of money in the account after 4 years

8. A store has used the equation $p = 30c - 19$ to approximate its profit in dollars on a given day when it has c customers. This year, it has adjusted the equation so c is replaced by $c - 1$. What effect does this have on the store's expected profit for this year?

A) Profits are expected to decrease by $3 per customer.
B) Profits are expected to decrease by $1 per customer.
C) The number of customers on a given day is expected to decrease by 1.
D) The number of customers on a given day is expected to increase by 3.

9. A meteorologist uses the equation $f(d) = -(d-3)^2 + 20$ to model the approximate amount of rain in inches expected to fall d days after the 1st day of the rainy season. Which statement is the best interpretation of the y-intercept of the graph of $y = f(d)$ in the xy-plane in this context?

A) On the 2nd day of the rainy season, 11 inches of rain are expected to fall.
B) On the 1st day of the rainy season, 11 inches of rain are expected to fall.
C) 3 inches of rain are expected to fall on the 20th day after the 1st day of the rainy season.
D) 20 inches of rain are expected to fall on the 3rd day after the first day of the rainy season.

10. The equation $y = .01x^3 + 1$ is used to show the relationship between the temperature of liquid x and liquid y in degrees Celsius in a solution used in an experiment.

A) When the temperature of liquid x is 0, the temperature of liquid y is $-\sqrt[3]{100}$.
B) When the temperature of liquid y is 0, the temperature of liquid x is $-\sqrt[3]{100}$.
C) When the temperature of liquid x is 0, the temperature of liquid y is 1.
D) When the temperature of liquid y is 0, the temperature of liquid x is 1.

Day 10

Drill Explanations

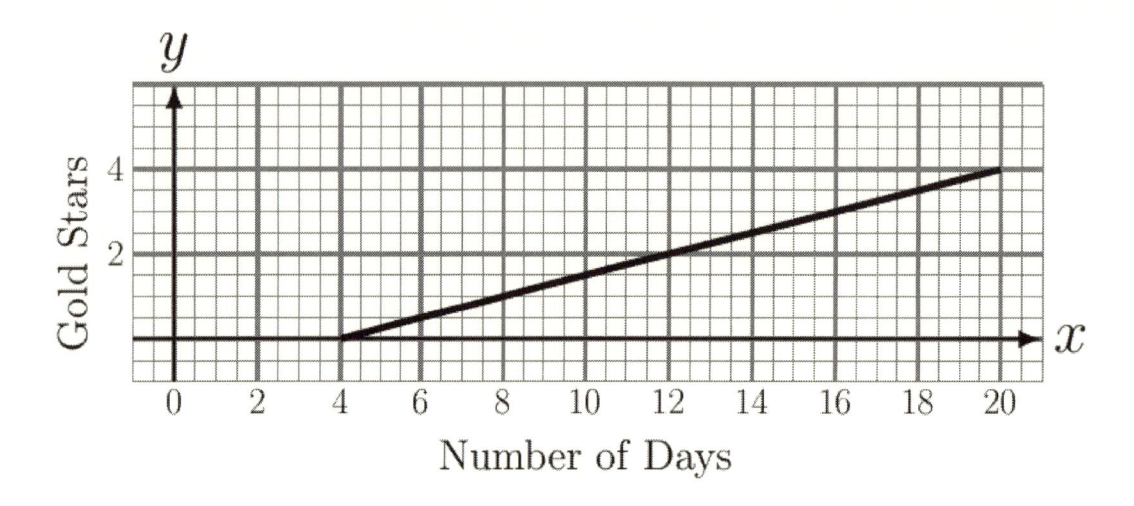

1. A 3rd grade teacher has devised a system for rewarding her students. Students who get all their work done for at least 8 consecutive days are awarded gold stars. The teacher then created the graph above to show the relationship between the number of days and the number of gold stars. Which of the following could be the equation representing this graph?

A) $y = \frac{4}{x} + 1$
B) $y = \frac{x}{4} - 1$
C) $y = \frac{x}{4} + 1$
D) $y = \frac{x}{4}$

It's worth noting that the y-intercept is not shown on the graph, which only shows x values starting with 4. We do see, however, that the points $(4, 0), (8, 1), (12, 2)$ and others appear to lie on the line. If you try these out in the equations, you'll see they only work in **choice B**. You could also use these points to determine the slope, which is ¼, and the y-intercept, which is -1. Again, note that this y-intercept is not visible since we don't see the actual y-axis.

Note: The next 3 questions all relate to the situation in Question 2.

2. An economic analyst has been tracking the performance of the price of 1 share of a company's stock. They use the equation $f(x) = (x - 3)^2 + 9.50$ to approximate this price x weeks from the coming Monday. What is the best interpretation of the y-intercept of the graph of $y = f(x)$ in the xy-plane in this context?

130 |Day 10

A) The price of 1 share of the stock this coming Monday will be $9.50.
B) In 3 weeks, the price of 1 share of the stock will be $9.50.
C) In 3 weeks, the price of 1 share of the stock will be $18.50.
D) The price of 1 share of the stock this coming Monday will be $18.50.

The *y*-intercept occurs where *x* is 0. You could plug 0 into the equation to get $y = 18.50$. This means that when *x*, which is the number of weeks **after** Monday, is 0, then *y*, which is the price of 1 share, is $18.50 on Monday. This is **choice D**.

3. What is the best interpretation of the coordinates of the vertex in this same context?

A)   In 3 weeks from the coming Monday, 1 share of the stock will reach its minimum price of $9.50.
B)   In 3 weeks from the coming Monday, 1 share of the stock will reach its minimum price of $18.50.
C)   In 3 weeks from the coming Monday, 1 share of the stock will reach its maximum price of $9.50.
D)   In 3 weeks from the coming Monday, 1 share of the stock will reach its maximum price of $18.50.

The equation is given in vertex form, so we know at a glance that the vertex is $(3, 9.50)$. Since this is an upward opening parabola, this is the minimum value: when *x* is 3 (weeks), then *y* (the price) is $9.50. This is **choice A**.

4. Why is there no *x*-intercept in this same context?

A) The equation does not show the price of 1 share of the stock on this coming Monday.
B) The equation does not show the current price of 1 share of the stock.
C) The price of 1 share of the stock will never reach a maximum value.
D) The price of 1 share of the stock will not reach 0.

An *x*-intercept occurs where *y* is 0. Algebraically, there is no real number solution for this because $(x-3)^2$ cannot equal -9.50. Also, if you graphed this on your OSC, you'll see that the parabola does not intersect the *x*-axis. Since *y* cannot equal 0, the price cannot be 0. This is **choice D**.

5. Phil typically drove the 250 miles to his cousin's home at an average rate of *r* mph. Because of improved road conditions, his average time making the trip most recently was $r + 6$ mph. Which of the following expressions represents how many fewer hours it took Phil to make the trip most recently?

A) $\frac{250}{r} - \frac{250}{r+6}$

B) $\frac{250}{r+6} - \frac{250}{r}$

C) $\frac{r}{250} - \frac{r+6}{250}$

D) $250r - 250(r-6)$

We'll use the rate formula in the form time = distance/rate. His typical time is therefore $\frac{250}{r}$. His recent time was $\frac{250}{r+6}$. This is **less** time. To get the difference, we subtract this from the **longer** time, as shown in **choice A**.

Drill Explanations | 131

6. Over 2 days, shoppers at a mall were asked to take part in a survey where they could rate how they felt about a soft drink on a scale of 1 to 12. On the 1st day, n people gave the drink an average score of s. The following day, $n - 10$ people gave the drink an average score of $s - 1$. If the sum of all the scores is 430, which of the following equations represents this situation?

A) $12ns + 10(n - 10) = 430$
B) $n(s - 1) + (n - 10)s = 430$
C) $ns + (n - 10)(s - 1) = 430$
D) $10ns + 10(ns - 1) = 430$

We'll use the average formula in the form average x number of terms = sum. On the 1st day, that gives us ns. On the 2nd day, that gives us $(n - 10)(s - 1)$. If we add these 2 sums together, we should get 430, as shown in **choice C**.

7. Four years ago, Saheed deposited x dollars into a savings account that generates compound interest biannually (twice a year). The expression $f(x) = (x)(1.0425^{2t})$ represents the amount in the account after a period during which no money was withdrawn or deposited. Which of the following is the best interpretation of t in this expression?

A) The amount that Saheed first deposited into the account
B) The number of years during which the account generates interest
C) The amount of money in the account after 2 years
D) The amount of money in the account after 4 years

We are told that Saheed deposited x dollars, so (1.0425^{2t}) must represent what this amount is multiplied by to get the amount after interest. The 1 represents the amount already in the account when the interest is calculated, and the .0425 (or 4.25%) represents the interest as a percentage of that amount. The $2t$ represents time, which is to say how many times the interest is calculated. Since we're told it's compounded twice a year, t represents the number of years, leading us to **choice B**.

8. A store has used the equation $p = 30c - 19$ to approximate its profit in dollars on a given day when it has c customers. This year, it has adjusted the equation so c is replaced by $c - 1$. What effect does this have on the store's expected profit for this year?

A) Profits are expected to decrease by $30 per customer.
B) Profits are expected to decrease by $30 per day.
C) The number of customers on a given day is expected to decrease by 10.
D) The number of customers on a given day is expected to increase by 30.

One approach is to pick a number for c. For example, if c equals 2, then using the original equation, we get a profit of 41. If we now subtract 1 from c, giving us $p = 30(2 - 1) - 19$, we get 11. Our profit decreased by 30 with the same number of customers. That leads us to **choice B**.

If we just do the algebra, we see that the new $p = 30(c - 1) - 19$ gives us $p = 30c - 30 - 19$, which simplifies to $p = 30c - 49$. This is 30 less than the original equation.

9. A meteorologist uses the equation $f(d) = -(d-3)^2 + 20$ to model the approximate amount of rain in inches expected to fall d days after the 1st day of the rainy season. Which statement is the best interpretation of the y-intercept of the graph of $y = f(d)$ in the xy-plane in this context?

A) On the 2nd day of the rainy season, 11 inches of rain are expected to fall.
B) On the 1st day of the rainy season, 11 inches of rain are expected to fall.
C) 3 inches of rain are expected to fall on the 20th day after the 1st day of the rainy season.
D) 20 inches of rain are expected to fall on the 3rd day after the 1st day of the rainy season.

The y-intercept occurs where x is 0. (In this case, we have d instead of x.) When we plug in 0 for d, we get $y = f(d) = 11$. Therefore, when **no** days have gone by after the 1st day of the rainy season—in other words, on the **1st** day—the amount of rainfall expected is 11 inches. This is **choice B**.

10. The equation $y = .01x^3 + 1$ is used to show the relationship between the temperature of liquid x and liquid y in degrees Celsius in a solution used in an experiment. Which statement is the best interpretation of the x-intercept of the graph of $y = f(d)$ in the xy-plane in this context?

The x-intercept occurs where y is 0. In this situation, that means the temperature of liquid y is 0. If we plug in 0 for y, we get $0 = .01x^3 + 1$, which simplifies to $-.01x^3 = 1$ and then $x^3 = -100$. If we take the cubed root of each side, we get $x = -\sqrt[3]{100}$. That leads us to **choice B**.

A) When the temperature of liquid x is 0, the temperature of liquid y is $-\sqrt[3]{100}$.
B) When the temperature of liquid y is 0, the temperature of liquid x is $-\sqrt[3]{100}$.
C) When the temperature of liquid x is 0, the temperature of liquid y is 1.
D) When the temperature of liquid y is 0, the temperature of liquid x is 1.

Day 11

Trigonometry and Function Notation – SOH CAH TOA Radians

After reading this material and completing the drill at the end, sign and check the box for Day 11.

Trigonometry

Here's the nice thing about trigonometry. There's not much of it on the SAT. You'll need to master a few things, and then you'll be able to efficiently handle the few trig questions you'll see. And it's likely that you already have had some exposure to SOH CAH TOA.

SOH CAH TOA

SOH CAH TOA is a handy way to remember how to handle the 3 main trig ratios: sine, cosine, and tangent.

The **s**ine of an angle is equal to the **o**pposite side over the **h**ypotenuse.
The **c**osine of an angle is equal to the **a**djacent side over the **h**ypotenuse.
The **t**angent of an angle is equal to the **o**pposite side over the **a**djacent side. (Notice that this is the one that doesn't involve the hypotenuse.)

The trig that you'll see on the test is all *right-angle trigonometry*, so let's look at that.

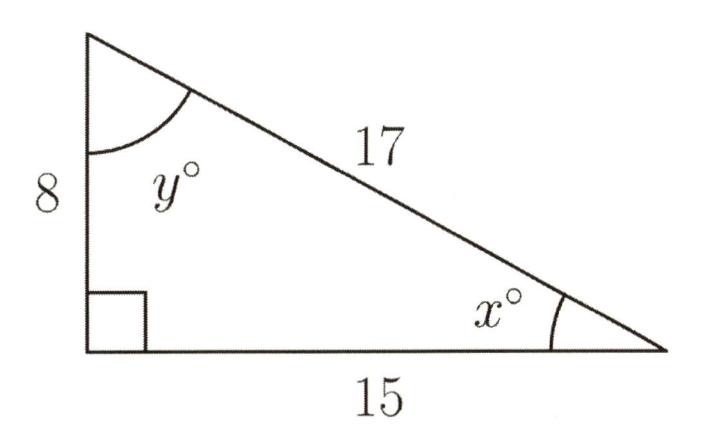

For this triangle, we know the following:

The sine of $x°$ equals $\frac{8}{17}$
The cosine of $x°$ is $\frac{15}{17}$
The tangent of $x° = \frac{8}{15}$

To clarify, the side *opposite* an angle is the side that does not create the angle, while the side *adjacent* to an angle is one of the sides that does create the angle.

As we can see, these trig ratios are simply fractions—one number over (divided by) another. Can these fractions be greater than 1? For sine and cosine, the bottom of the fraction is the denominator, and since that can't be smaller than the numerator (it's the biggest of the 3 sides), the fraction cannot be greater than 1. The sine of an angle can, however, **equal** 1. The sine of 90 degrees is hypotenuse over hypotenuse, which equals 1.

On the other hand, the tangent of an angle can be whatever it wants, including, for example, the "fraction" $\frac{2}{1}$, which, of course, equals 2.

We can use the triangle above to discover a general rule. Notice that the sine of x is the same as the cosine of y; they're both $\frac{8}{17}$. In other words, for x, opposite over hypotenuse is the same as adjacent over hypotenuse is for y, and vice versa. And whatever x and y are, their sum has to be 90 degrees since, like all triangles, the sum of all 3 angles is 180 degrees. Putting this together, we can say that if 2 angles add up to 90 degrees, the sine of one of them equals the cosine of the other (in both directions).

Let's try a question.

One leg of right triangle ABC is 5 inches long, and the hypotenuse is 10 inches long. What is the cosine of the smallest angle in the triangle?

Once you know we're dealing with a right triangle, you should start thinking about the special right triangles. And in fact, this is one of those. The only right triangle with a hypotenuse twice the length of 1 of the legs is the 30/60. Remember? Its sides are in the proportion $1 : \sqrt{3} : 2$. That means the other leg (the longer one) is $5\sqrt{3}$ inches long.

Trigonometry and Function Notation – SOH CAH TOA Radians | 135

Okay, now we need to identify which angle is the smallest. By definition, this **must** be the one opposite the shortest leg. Since the shortest leg is 5, we want the cosine of the angle across from 5. We have this situation.

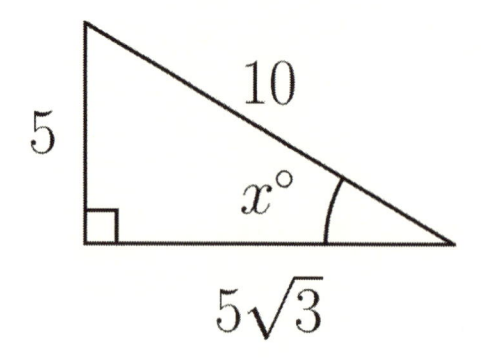

```
         10
    5   x°
      5√3
```

We want the cosine of x, which is adjacent over the hypotenuse, and that would be $\frac{5\sqrt{3}}{10}$. This can be simplified to $\frac{\sqrt{3}}{2}$.

If you haven't already, you'll want to know SOH CAH TOA without having to think about it. And some trig ratios are worth memorizing. We've already said that the sine of 90 degrees is 1. You should also memorize that the sine of 30 degrees is ½.

Please remember that you can only find the sine, cosine, or tangent (often abbreviated sin, cos, tan) of an **angle**. There's no such thing as finding, for example, the tangent of a side.

Speaking about tangents, **the tangent of an angle is equal to the sine of that angle divided by the cosine of that angle**. This makes sense if you think about it. $\frac{sine}{cosine} = \frac{\frac{opposite}{hypotenuse}}{\frac{adjacent}{hypotenuse}} = \frac{opposite}{adjacent} = tangent$.

Radians: Another Way to Measure an Angle

Another trig term to know is *radians*. That is simply a different way to measure angles. You could use degrees, or you could use radians. For example, we could say that an angle is 100°, or we could say that it is $\frac{5\pi}{9}$ radians. There's no difference. Of course, these measurements look different. When using radians, we'll always see π in the value.

You should be comfortable converting from degrees to radians. It helps if you just remember that **180 degrees is equal to π radians**. From there, it's fairly obvious that 360 degrees is equal to 2π radians, and 90 degrees is equal to $\frac{\pi}{2}$ radians. You can think of it this way: Divide the number of degrees by 180, and then multiply that by π.

Let's try this out.

One angle of a right triangle is 50°. What is the measure of the other non-right angle expressed in *radians*?

The other angle must equal 40 in order for the sum to be 180. Therefore, all we need to do is convert 40° to radians. We'll divide that by 180, giving us $\frac{2}{9}$. Multiplying this by π, we get $\frac{2\pi}{9}$ radians. (Note that we want a fraction, not a decimal, and that we leave π as π.) Note: Calculators, including the OSC, allow you to toggle between degrees and radians. Know which one is turned on.

Function Notation: f(x)

We've already encountered function notation when looking at graphs where $f(x) = y$. (This would be pronounced "the f of x equals y.") And we've seen function notation in equations such as $g(x) = x^2 - 2$.

Sometimes we'll see this kind of notation without any graphs involved. A function is an instruction. If you're told that $f(x) = 3x - 2$, that means any x value should be multiplied by 3, and then 2 should be subtracted from that. So if $x = 5$, you would say $f(5) = 3(5) - 2 = 13$—in short, $f(5) = 13$.

We also see *nested* functions such as $f[g(x)]$, which is pronounced "the f of the g of x." We always work from the *innermost* function out, so you would handle $g(x)$ first. For example, let's say that $g(x) = x^2 + 1$, and it's still true that $f(x) = 3x - 2$. For $f[g(3)]$, you would first determine that $g(3) = 3^2 + 1 = 10$. Now that you know that $g(3) = 10$, you can determine that $f(10) = 3(10) - 2 = $ **28**.

Sometimes the test will **give** you the final result of the function and ask you to determine what the missing value is. Here's a question of that sort.

The function *f* is defined by $f(x) = x^2 - 50$. For what value or values of x does $f(x) = -25$?

A) −50
B) −25 and −5
C) −5 and 5
D) 5

One simple way to solve this is to try out the choices. If you plug each one in for x in the function equation, you'll see that the answer is **C**. You could also solve it algebraically by setting up the equation $x^2 - 50 = -25$. From here, you can add 50 to each side, but be careful. When $x^2 = 25$, there are 2 solutions, both positive and negative 5.

Day 11

Drill

Complete all the questions below.

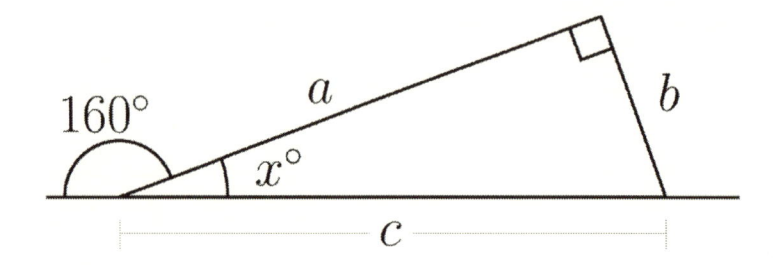

1. In the triangle above, *a, b,* and *c* are the lengths of the sides. Which of the following must be true?

A) $Sin\ 20° = \frac{a}{b}$
B) $Cos\ 20° = \frac{a}{b}$
C) $Sin\ 20° = \frac{b}{c}$
D) $Cos\ 20° = \frac{b}{c}$

2. Triangles ABC and DEF are similar. Side *a* corresponds to side *d*, and side *b* corresponds to side *e*. Also, side *a* is across from angle *A*, and so on. If $tan\ \square ABC = \frac{12}{5}$ and $cos\ \square ABC = \frac{5}{13}$, what is $sin\ \square DEF$?

3. If $sin(90 - x) = cos\ y$, which of the following must be true?

A) $x = y$
B) $sin\ x = cos\ y$
C) $sin\ x + cos\ y = 90$
D) $sin\ x - cos\ y = 90$

4. If the tangent of $\square RST \approx 0.29$ and the cosine of $\square RST \approx 0.96$, which of the following is the best approximation of the sine of $\square RST$?

A) 0.28

138 |Day 11

B) 0.67
C) 1.25
D) 3.3

5.
$$g(x) = x^2 - x$$

For the function g defined as shown, what is the value of $\frac{g(7)}{g(-2)}$?

6.
$$f(x) = -x$$
$$g(x) = \frac{3}{x}$$

For the functions f and g defined as shown, what is the value of $f[g(-9)]$?

7. The function $t(x) = 5x - 2.5$ is used to approximate the temperature in degrees Celsius of a solution x seconds after the start of an experiment. Using this function, how many seconds have elapsed when the temperature reaches 15 degrees Celsius?

8.
$$f(x) = x^2 - 3x + 2$$

For the function f defined above, which of the following equations would make the expression $\frac{x-1}{x-2}$ undefined?

A) $f(x) = -1$
B) $f(x) = 0$
C) $f(x) = 1$
D) $f(x) = 2$

Drill | 139

Day 11

Drill Explanations

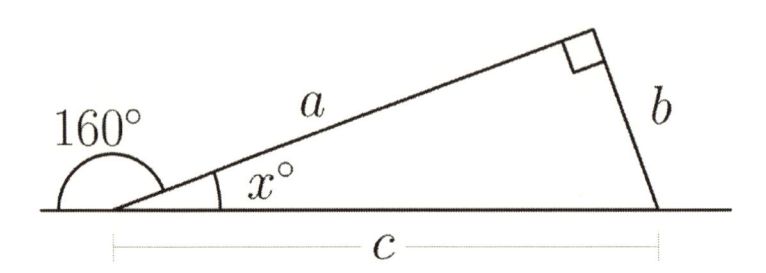

1. In the triangle above, *a, b,* and *c* are the lengths of the sides. Which of the following must be true?

A) $Sin\ 20° = \frac{a}{b}$
B) $Cos\ 20° = \frac{a}{b}$
C) $Sin\ 20° = \frac{b}{c}$
D) $Cos\ 20° = \frac{b}{c}$

We know that $x = 20$ because straight lines have 180 degrees. Going through the choices, only choice C gives us a trig ratio correct for this triangle. The sine of an angle is Opposite/Hypotenuse, which in this case is *b/c*.

2. Triangles ABC and DEF are similar. Side *a* corresponds to side *d*, and side *b* corresponds to side *e*. Also, side *a* is across from angle *A*, and so on. If $tan\ \square ABC = \frac{12}{5}$ and $cos\ \square ABC = \frac{5}{13}$, what is $sin\ \square DEF$?

Since the triangles are similar, they have the same angles. That means the sine, cosine, and tangent for the corresponding angles are the same. Because we are told which sides correspond, we know that $\square ABC = \square DEF$. Therefore, the sine of angle *DEF* will be the same as the sine of *ABC*. Here's a diagram of the situation, which applies to either triangle. (Note that this is a special triangle, the 5, 12, 13 right triangle. This does not mean that the sides are those exact lengths but are rather in that ratio. For example, the sides could actually be 50, 120, and 130. That's why the sides are marked as 5x, etc.)

140 |Day 11

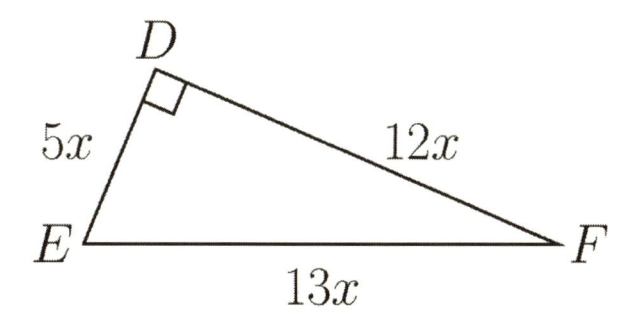

We can see now see that the sine of *DEF* (as well as *ABC*) is equal to 12x/13x, or simply $\frac{12}{13}$.

3. If sin(90 − x) = cos y, which of the following must be true?

A) $x = y$
B) $\sin x = \cos y$
C) $\sin x + \cos y = 90$
D) $\sin x - \cos y = 90$

This question is based on the rule that if 2 angles sum to 90 degrees, the sine of 1 of them equals the cosine of the other, in both directions. We could also express it this way. If the sine of 1 angle equals the cosine of the other, they sum to 90 degrees. Given the original equation, we can say that $(90 - x) + y = 90$. We can simplify this to $90 - x + y = 90$, and after subtracting 90 from both sides and then adding x to both sides, we get $x = y$, **choice A**.

4. If the tangent of □*RST* ≈ 0.29 and the cosine of □*RST* ≈ 0.96, which of the following is the best approximation of the sine of □*RST*?

A) 0.28
B) 0.67
C) 1.25
D) 3.3

You might first recognize that choices C and D can be eliminated because they are both greater than 1, and no sine (or cosine) can be greater than 1. You also should keep in mind that the tangent of an angle is equal to its sine divided by its cosine. Expressed another way, the sine of an angle is equal to its tangent times its cosine. In this case, we get 0.29×0.96, which gives us approximately 0.28, which is **choice A**.

5.
$$g(x) = x^2 - x$$

For the function *g* defined as shown, what is the value of $\frac{g(7)}{g(-2)}$?

All we need to do here is separately calculate the numerator, then the denominator, and then divide. So we have $(7) = 7^2 - 7 = 42$ and $g(-2) = (-2)^2 - (-2) = 6$. We can now divide, giving us $\frac{42}{6} = 7$.

Drill Explanations | 141

6.
$$f(x) = -x$$
$$g(x) = \frac{3}{x}$$

For the functions f and g defined as shown, what is the value of $f[g(-9)]$?

We'll start with $g(-9)$, which is equal to $\frac{3}{-9} = -\frac{1}{3}$. Now we can take $f\left(-\frac{1}{3}\right)$, which is equal to $-\left(-\frac{1}{3}\right)$, which is $\frac{1}{3}$.

7. The function $t(x) = 5x - 2.5$ is used to approximate the temperature in degrees Celsius of a solution x seconds after the start of an experiment. Using this function, how many seconds have elapsed when the temperature reaches 15 degrees Celsius?

We are looking for x, so we can set up the following equation: $15 = 5x - 2.5$. After some simple algebraic manipulation, we get $x = 3.5$.

8.
$$f(x) = x^2 - 3x + 2$$

For the function f defined above, which of the following equations would make the expression $\frac{x-1}{x-2}$ undefined?

A) $f(x) = -1$
B) $f(x) = 0$
C) $f(x) = 1$
D) $f(x) = 2$

An expression is undefined when the denominator is 0. If we try the choices, we see that **choice B** gives us $0 = x^2 - 3x + 2$. If we factor, we see that $0 = (x-2)(x-1)$, which means that the solutions are $x = 2$ and $x = 1$. When $x = 2$, the denominator is 0, and that makes the expression undefined.

Day 12

Graphics – Probability, Mode, Median, Probability I, Standard Deviation

---

After reading this material and completing the drill at the end, sign and check the box for Day 12.

The *xy*-coordinate planes we've worked with will not be the only graphics to appear on the test. You've already seen most, if not all, of these other graphs in school, but we'll have a closer look at them here. The actual math tested this way can vary from algebra to probability. We'll cover everything.

Tables

Tables list values, sometimes including unknown values such as *x*. These questions often involve simple algebra.

x	y
-1	-14
1	-4
2	1
3	6

The table shows 4 values of *x* and their corresponding values of *y*. Which of the following equations represents the linear relationship between *x* and *y*?

A) $y = x - 13$
B) $y = 14x$
C) $y = x - 1$
D) $y = 5x - 9$

The numbers in the table and the equations are simple enough that you just might want to try the numbers in each choice. For example, while the 1st pair of values (-1, -14) fit both choice A and choice B, only choice D works for every pair. (Trying 3 pairs is usually a good idea.) Instead, you could use the values to determine the slope. As always, that's the difference in the *y* values **over** the difference in the *x* values of any 2 points on the line. Whichever 2 you choose, you'll get 5, making **choice D** correct.

Here's a similar question involving a quadratic.

x	f(x)
-2	13
-1	7
3	23

The table shows 4 values of *x* and their corresponding values of *f(x)*. If a quadratic equation representing this relationship is given in the form $f(x) = ax^2 + c$ where *a* and *c* are constants, what is the value of *a*?

We'll start by again plugging the values into the equation. If we start with the 1st pair, we get $13 = a(-2^2) + c$, which simplifies to $\underline{13 = 4a + c}$. We still have 2 unknowns, so we'll try the next pair as well, giving us $7 = a(-1^2) + c$, which simplifies to $\underline{7 = a + c}$. Now that we have 2 equations, we can solve for the 2 unknowns through substitution or elimination. If, for example, we subtract the second equation from the 1st one, we get $6 = 3a$, which means that ***a* = 2**.

Probability

We don't see probability tested much on the SAT, but it can show up, for example, in a table-based question such as this one.

The given table shows the distribution of the 1st, 2nd, and 3rd year students who are majoring in pre-medicine, electrical engineering, or graphic design at a certain college.

	1st year	2nd year	3rd year
Pre-med	35	16	11
Elec. Eng.	32	18	11
Graphic Des.	38	20	14

If 1 of these students is selected at random, what is the probability of selecting a student majoring in electrical engineering, given that the student is a 2nd year student?

Though probability is often expressed as a percentage, it's usually given as a fraction on the test. The bottom of the fraction will be the **total** number of possible outcomes. The numerator is the number of possible things we **desire** to select. In this case, the total is the sum of all the 2nd year students. (The others cannot be selected because of the way the question is phrased.) The numerator is the number of 2nd year electrical engineering students. That means the fraction is 18/54, which can be simplified to $\frac{1}{3}$.

Sometimes we see tables used in samples and polls.

<table>
<tr><td colspan="2" align="center">Poll Results for Prop 2</td></tr>
<tr><td>For</td><td>210</td></tr>
<tr><td>Against</td><td>n</td></tr>
</table>

The table above shows the results of a poll where a sample of voters were asked if they supported an upcoming proposition they could vote for or against in an upcoming election. After the election, in which 28,793 people voted, it was determined that 12,340 people voted "for," which was quite close to the expected number given the results of the poll. Which of the following is the most likely number of people who responded "against" in the poll?

A) 218
B) 280
C) 560
D) 4,113

Like many sample questions, this can be handled as a proportion/ratio. We can start by first determining how many people actually voted "against." Since there were 28,793 voted in total, we can subtract 12,340 to get 16,453 "against" votes. Now, we can set up the proportion with an equation: $\frac{210}{n} = \frac{12,340}{16,453}$. To get n, we could use our OSC to cross-multiply. You should get a value for n quite close to 280, which is **choice B**.

Note: Do not put commas in the numbers on the OSC because the calculator thinks you're writing, for example, the coordinates (12, 340).

Other Graphics

Bar graphs and histograms often appear on the exam. These look similar, but a histogram such as the one below shows continuous values, as can be seen by the lack of any gaps between the bars. In this figure, the 1st bar represents values greater than or equal to 0 but less than 5. The next bar represents values greater than or equal to 5 but less than 10.

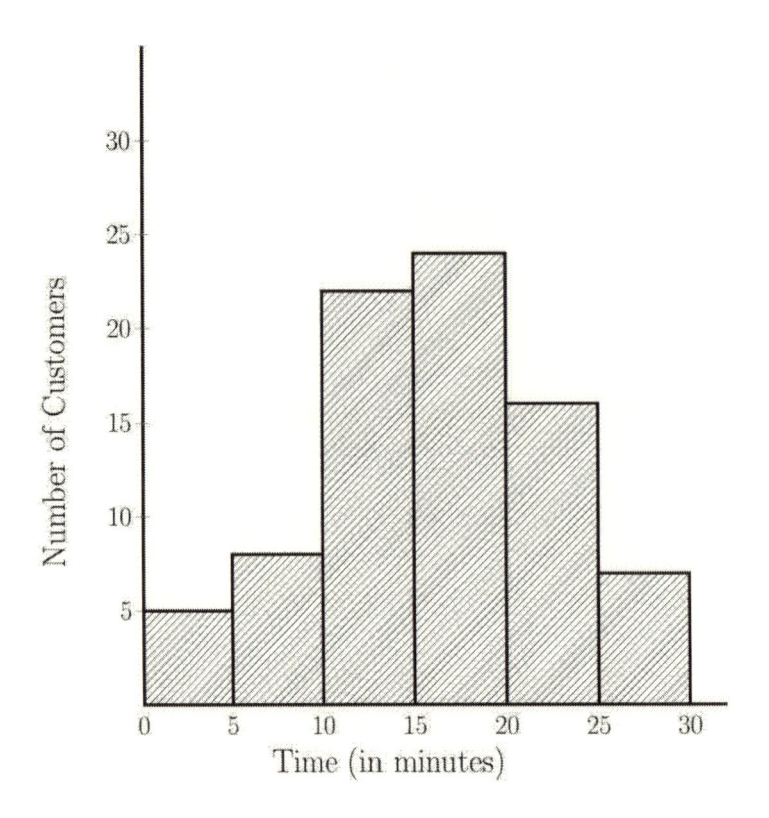

The questions attached to these graphs sometimes involve the concepts of mean (average) and **median**. Median is the middle value of a set of values **if** there is an odd number of values. If the set contains an even number of values, then there is no middle value, and the median is **the average of the 2 center values**. So in the set consisting of these values (-1, 5, 12, 14, 50, 99), the median is 13 because that's the average of 12 and 14.

Here's a question using this graph.

The histogram shown summarizes the amount of time spent in a store one day by the 82 customers who entered the store. Which of the following could be the median amount of time in minutes that a customer spent in the store that day?

A) 7
B) 15
C) 22
D) 30

We're told there are 82 customers. That means the median time would be the average of the time spent by the 41st and 42nd customers if we were to line up each person in ascending (or descending) order. For example, maybe the person who spent the least amount of time (we'll call that person ONE) spent 2 minutes in the store. Person TWO spent 3.5 minutes in the store, and so forth.

We know that the first 5 people all spent no more than 5 minutes in the store. Moving up, another 8 people (our best estimate) are in the next group, so that takes care of the first 13 people. Then we have another 22 people (or so), the people who spent at least 10 minutes but not more than 15 minutes in the store. That's 5 + 8 + 22 = 35 people so far. The next group has close to 25 people in it, so this group would contain the 41st and 42nd customers, and their average, whatever it is, would be a number that was in this group, which is the group that spent at least 15 minutes and less than 20 minutes in the store. That means the answer is choice B, the only number in that range.

Another concept that might be tested is **mode**. This is simply the value (if any) that shows up the most. If no value shows up more than each of the others, there's no mode. Here's a **dot graph** testing mode.

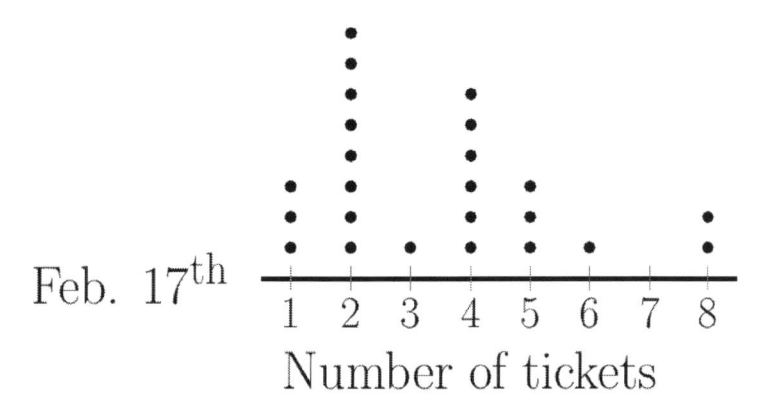

Feb. 17th

Number of tickets

The dot plot shows the distributions of the number of tickets sold each time a purchase was made for a concert to be held on February 17th. What is the mode of this data set?

Each dot represents the number of tickets purchased of a certain quantity. For example, there were 3 puchases of 1 ticket and no purchases of 7 tickets. It's obvious that most of the purchases were of 2 tickets, so the mode is **2**. Note: The exact number of tickets sold (in this case, 83) is not important and is **not** the answer.

Box plots—sometimes called box-and-whisker plots—are somewhat rare but can appear on the test. They are created from 5 numbers—the median, the minimum, the maximum, the median of all the numbers *less* than the median of all the numbers, and the median of all the numbers *more* than the median of all the numbers. Here's an example.

Let's say that during a month, a real estate company sold 7 houses. The selling prices of the houses, *in thousands*, were 205, 310, 425, 415, 170, 395, and 319. First, let's get the median for the whole set. If you arrange them in order, you'll get 170, 205, 310, 319, 395, 415, and 425. With 7 values, the median is the middle one—319.

The minimum and maximum are 170 and 425.

The numbers less than the median are 170, 205, and 310. The median of these is 205.

The numbers more than the median are 395, 415, and 425. The median of these is 415.

Now we have the 5 numbers we wanted. Here's what the box plot would look like.

Graphics – Probability, Mode, Median, Probability I, Standard Deviation | 147

                    150   200   250   300   350   400   450
                              Price (in thousands)

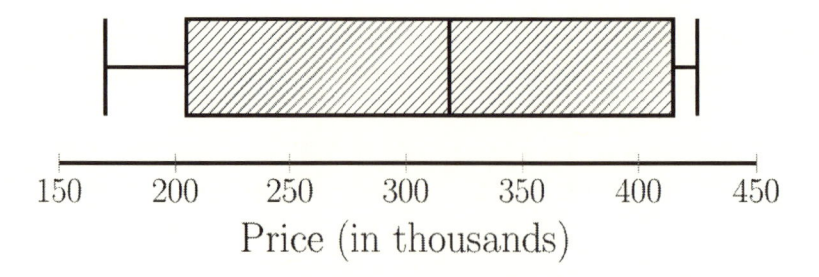

The outer horizontal lines on each side extend to the minimum and maximum values. The center vertical line indicates the value of the median. The 2 vertical lines on either side of the center one indicate the lower median and the upper one. (These are sometimes called *quartiles*.)

Here's a question that might accompany this figure.

The box plot summarizes the selling price (in thousands of dollars) of 7 houses that were sold by a real estate company during a month. None of the houses sold for the same price. Which of the following *could* be true?

A) The selling price of exactly 2 of the houses was less than $200,000.
B) The selling price of exactly 2 of the houses was less than $300,000.
C) The selling price of exactly 4 of the houses was more than $350,000.
D) The difference between the selling price of 2 of the houses was $300,000.

We know the numbers because **we** created the graph, but let's just pretend that all we know is the figure. We can't tell the exact numbers, but we do know that 7 houses were sold, which means that 3 were sold for less than the median, which looks to be, in thousands, about 325, and 3 were sold for more than that.

We can eliminate choice A because from the graph we can see that only 1 house sold for less than $200,000, the house that had the minimum selling price. We can eliminate choice C because we know that only 3 were sold for more than the median, which is close to $325,000. We can eliminate choice D because the biggest difference in selling price would be between the minimum and maximum, and we can see that would be less than $300,000. The horizontal line*s* don't extend far enough.

That leaves **choice B**. That *could* be true. There's no reason that 2 of the houses couldn't have sold for less than $300,000.

Mode and median are part of the field of *statistics*, as is average. Below is a related concept.

Standard Deviation

Standard deviation can be time-consuming to calculate, but the SAT has never required that test-takers do this. However, you should understand the concept, and you should be able to **compare** the standard deviation (SD) of data sets.

SD refers to how spread out the values are from the mean (average). For example, let's say we have this set of values: 3.8, 3.9, 4, 4.05, 4.25. If we were to add these 5 numbers and divide by 5, we'd find the average is 4. You'll notice that relatively speaking, all the values in the set are close to 4. There isn't much spread from the average. But here's another set of numbers that **also** have an average of 4: -22, -5, 0, 2, 45. In this case, the numbers **are** fairly spread out from 4. This set, therefore, has a **larger** standard deviation.

Here's a pair of bar graphs attached to an SD question.

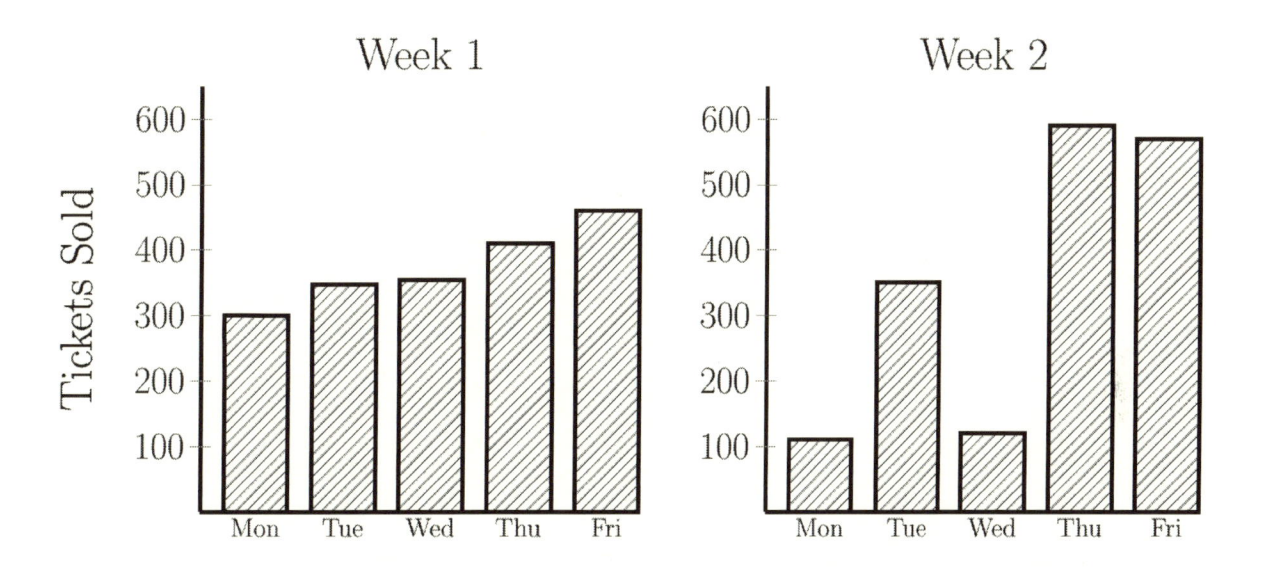

The bar graphs shown represent the expected number of tickets sold to a festival from Monday to Friday during week 1 and week 2 of a festival. Which of the following is true?

A) **The standard deviation of the number of tickets sold each day is greater during week 1 than during week 2.**
B) **The standard deviation of the number of tickets sold each day is greater during week 2 than during week 1.**
C) **The standard deviation of the number of tickets sold each day is the same during both weeks.**
D) **There is no way to determine which week had the greater standard deviation of the number of tickets sold each day.**

Though we can't determine exact values from these graphs, we can tell that they have similar averages. In week 1, there are 2 days close to or a little less than (about) 350 as well as 2 days that are a little more than 350. Friday's value is high enough that the average is probably somewhat more than 350. On week 2, we see 2 days that are quite a bit less than (about) 350 as well as 2 days that are quite a bit more than 350. This comparison tells us that the averages of the 2 weeks are close, **but** it's clear that the values in week 2 are much more spread out from the average, giving this one the greater SD, as it says in **choice B**.

Graphics – Probability, Mode, Median, Probability I, Standard Deviation | 149

Margin of Error

This refers to how far "off" an estimated or expected value could be from the actual value. For example, if 10 people are asked who they think will win an election in which 1 million will vote, that will have a larger margin of error than if 10,000 people were asked. Margins of error are typically given in percentage form and refer to how "off" an estimated number is **in both directions**—too big or too small. Here's a question with a **pie chart** that tests this idea.

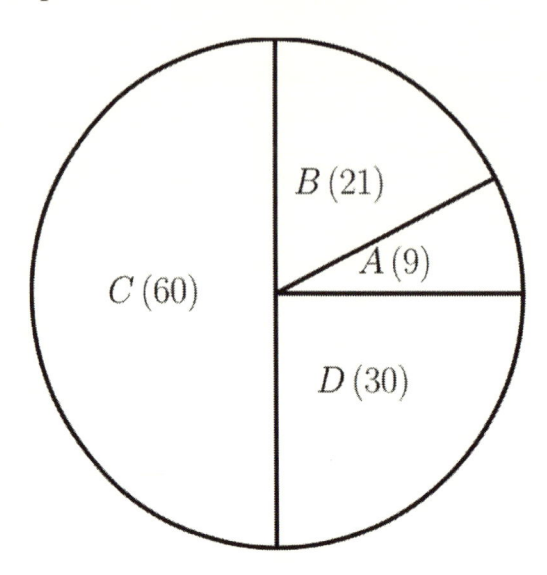

As part of a study, a group of 120 people were observed as they listened to 4 songs. The songs were labeled A, B, C, and D. Based on observations, a circle graph, shown above, was created to show which song each person seemed to prefer. The results were given a margin of error of 5%. Based on this, what is the largest possible difference between the number of people who preferred 2 different songs? (Round all values up to the nearest integer.)

The biggest difference shown is between the 9 people who preferred song B and the 60 who preferred song D. To get the largest possible difference, let's say that 9 is too high and 60 is too low. Let's start with 60. The margin of error is 5%, so we'll add that to 60. If we multiply 1.05 by 60, we get 63. Now we'll lower the 9 by 5%. If we do this by multiplying .95 x 9, it's 8.55. Since we're told to round up, that brings us back to 9. The difference is the difference between 63 and 9, which is **54**.

We'll finish with a return to our *xy*-plane and to the "non-real world" of pure math so we can look at the **line of best fit and scatterplots**. As you might know, a scatterplot is simply a bunch of plotted points, and the line of best fit is the line that most closely "fits" these points.

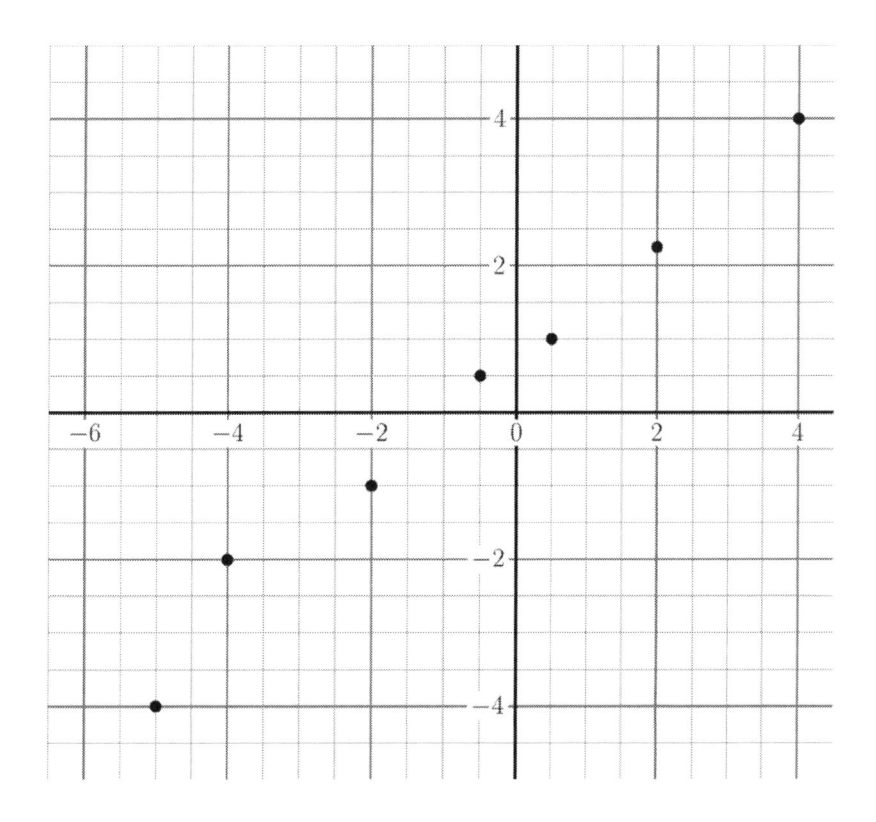

Which of the following equations is the most appropriate linear model for the data shown in the scatterplot?

A) $y = .8x + 1$
B) $y = x + 1.8$
C) $y = 2x + 1$
D) $y = 8x - 1$

As mentioned on an earlier day, it's a good idea to have some idea of what different slopes look like. For example, a line that intersects the origin (0, 0) and bisects quadrants I and III (the upper right and lower left quadrants) has a slope of 1. (This is assuming that the scale on each axis is the same, as it is here.) The line of best fit here would be pretty close to that, which means that choice C and definitely choice D are no good. The *y*-intercept looks closer to 1 than to 1.8, so that leads us to **choice A**.

Day 12

Drill

Complete all of the following questions.

1.

x	y
-1	0
0	-5
2	-13

The table shows 4 values of x and their corresponding values of y in a linear relationship. What is the slope of this relationship?

2.

x	f(x)
-12	267
-2	87
2	31
12	31

The table shows 4 values of x and their corresponding values of $f(x)$. What is the x-coordinate of the vertex of a parabola defined by this relationship?

3.

Lopez	Barnes	Cruz	Davids
32	36	15	40

152 |Day 12

The table shows which of 4 candidates were preferred by a sample of potential voters in an upcoming election in which about 21,000 people are expected to vote. This sample has an associated margin of error of 4%. Using this data, which of the following is the best approximation of the number of votes that Davids will receive in the election?

A) 4,160
B) 7,000
C) 7,280
D) 12,300

4.

	R	B	G
Small	10	7	12
Large	14	2	x

The table shows the number of small and large marbles, each of which is red, blue, or green, in a bag. If a large marble is pulled out of the bag at random, the probability that it is green is 68%. How many large green marbles are in the bag?

A) 16
B) 32
C) 34
D) 68

5.

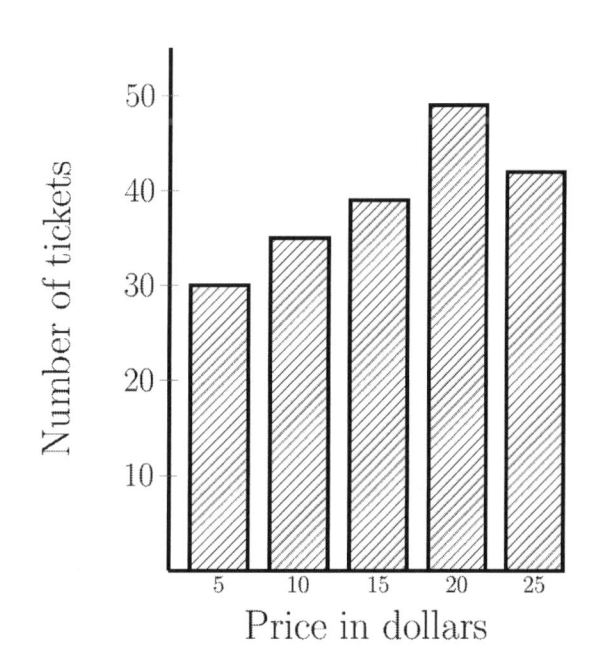

At a school fair, attendees could purchase a ticket for $5, $10, $15, $20, or $25. The more expensive the ticket, the more activities were included in the price. The bar graph shows the number of tickets purchased at each of these dollar amounts. Which of the following is the mode of the dollar price of the tickets sold?

A) 15
B) 20
C) 25
D) 48

6.

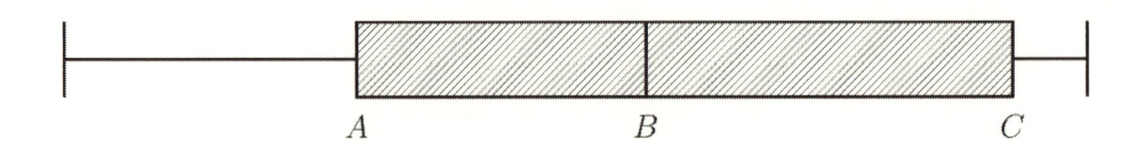

Set A consists of the following values: -10, -6, -3, -1, 3, and 4. The box plot shown is a graphic representation of this set. What is the sum of the values located at points A, B, and C?

7.

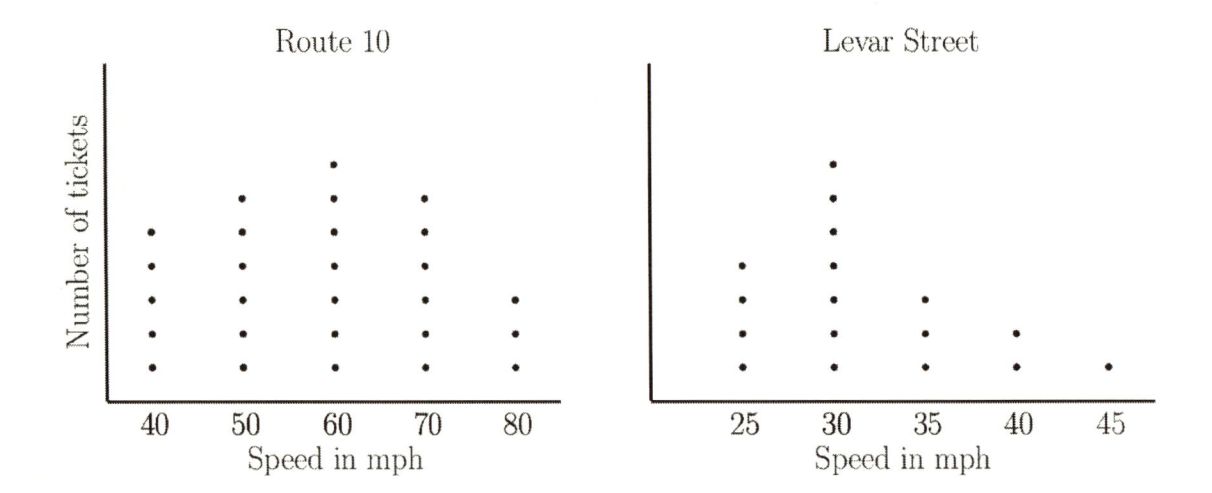

A record was kept of the speed of the drivers as they passed 2 different points, 1 on Route 10 and the other on Levar Street. The dot plot shown is a record of the approximate speed of the drivers as they passed these points. Which of the following is true?

A) The standard discrepancy of the speed of the Route 10 drivers is greater than the standard discrepancy of the speed of the Levar Street drivers.
B) The standard discrepancy of the speed of the Route 10 drivers is less than the standard discrepancy of the speed of the Levar Street drivers.
C) The standard discrepancy of the speed of the Route 10 drivers is equal to the standard discrepancy of the speed of the Levar Street drivers.
D) There is not enough information to compare the standard discrepancy of the speed of the drivers on Route 10 and the speed of the drivers on Levar Street.

8.

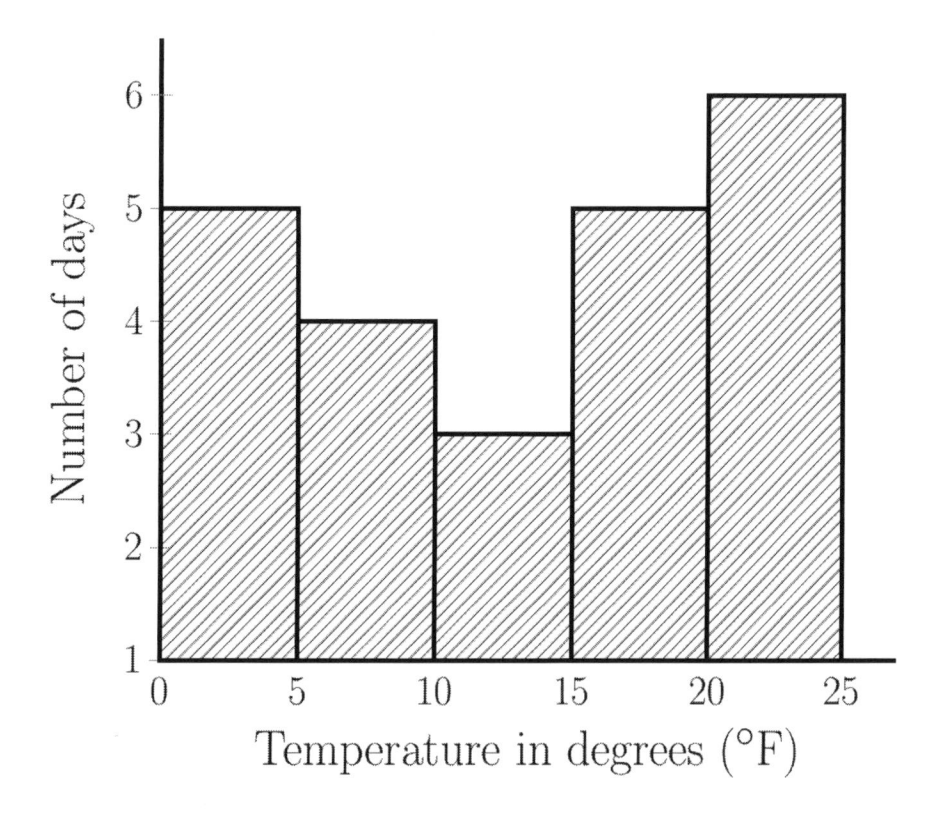

The average daily temperature at a certain location was recorded on 23 consecutive days. The histogram shown is a record of the coldest temperatures on these days. The 1st bar, for example, shows the number of days that the coldest temperature was at least 0 degrees and less than 5 degrees. The associated margin of error for the temperatures is 5%. Based on this graph, what was the coldest possible temperature for the day on which the median temperature was recorded?

9.

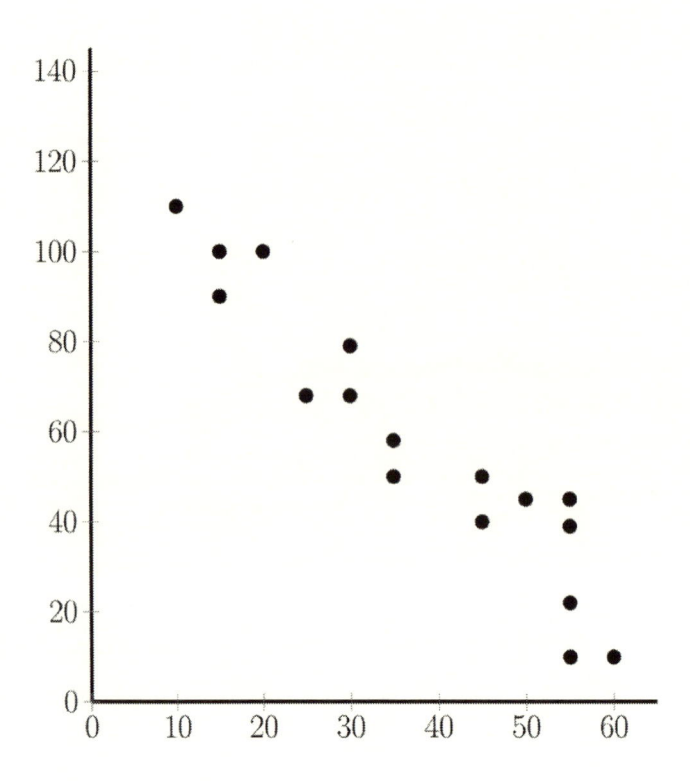

Which of the following equations is the most appropriate linear model for the data shown on the scatterplot?

A) $y = -x + 130$
B) $y = -\frac{11}{6}x + 60$
C) $y = \frac{11}{6}x + 130$
D) $y = -\frac{11}{6}x + 130$

Day 12

Drill Explanations

1.

x	y
-1	6
1	-4
2	-9

The table shows 4 values of x and their corresponding values of y in a linear relationship. What is the slope of this relationship?

We're told that these values are in a linear relationship, so to find the slope, we just need to set up a fraction with the difference between 2 y-values on top of the difference between 2 x-values. In other words, we want rise over run. If we use the 1st 2 pairs of values, we get $\frac{6-(-4)}{-1-1} = -\frac{10}{2} = -5$. Of course, this will work with any 2 sets of coordinates given in the table.

2.

x	f(x)
-12	267
-2	87
2	31
12	31

The table shows 4 values of x and their corresponding values of f(x). What is the x-coordinate of the vertex of a parabola defined by this relationship?

The simplest approach here is to notice that 31 is the $f(x)$ for 2 values—2 and 12. Because they share the same y-coordinate when graphed, the vertex of the parabola is midway between the points (2, 31) and (12, 31). This alone won't tell us what the y-coordinate of the vertex is, but we don't need that. The average of 2 and 12 is 7, so the vertex has an x-coordinate of 7.

3.

Lopez	Barnes	Cruz	Davids
32	36	15	40

The table shows which one of 4 candidates was preferred by a sample of potential voters in an upcoming election in which about 21,000 people are expected to vote. Using this data, which of the following is the best approximation of the number of votes that Davids will receive in the election?

A) 4,000
B) 7,000
C) 8,000
D) 12,000

Adding up the numbers in the table, we see that 123 people responded to the question about candidate preference. Of these, 40 selected Davids. Since 40 out of 123 is about 1/3, we expect about 1/3 of all the voters to select Davids. One-third of 21,000 is 7,000, making **B** the correct choice. Choice C isn't far off, but since $\frac{40}{123}$ is actually a little **less** than 1/3, the actual number of voters selecting Davids is likely to be a little less than 7,000.

4.

	R	B	G
Small	10	7	12
Large	14	2	x

The table shows the number of small and large marbles, each of which is red, blue, or green, in a bag. If a large marble is pulled out of the bag at random, the probability that it is green is 68%. How many large green marbles are in the bag?

A) 16
B) 32
C) 34
D) 68

We are told that a large marble is pulled out at random, so we have $16 + x$ as our **total number of possible marbles selected**. There are x green marbles, so we can say that the probability of selecting a green large marble from the total number of large marbles is $\frac{x}{16+x}$. It's given that this probability is 68%, so our equation is $\frac{x}{16+x} = 68\%$. Since we'll be cross-multiplying and solving for x, it's best to think of 68% as either .68 or $\frac{68}{100}$, which simplifies to $\frac{17}{25}$. If we use the decimal (.68) and cross-multiply, we get $x = .68(16) + .68x$. Combining like terms, we get $.32x = 10.88$ and then $x = 34$, **choice C**.

5.

158 |Day 12

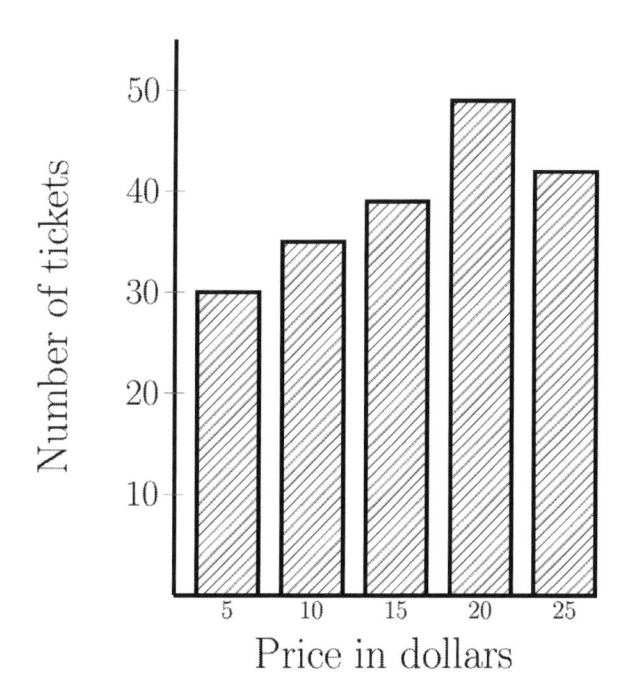

At a school fair, attendees could purchase a ticket for $5, $10, $15, $20, or $25. The more expensive the ticket, the more activities were included in the price. The bar graph shows the number of tickets purchased at each of these dollar amounts. Which of the following is the mode of the dollar price of the tickets sold?

A) 15
B) 20
C) 25
D) 48

More $20 tickets were sold than any other priced ticket. That makes **choice B** correct.

6.

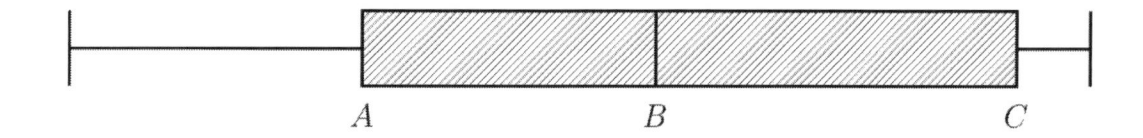

Set A consists of the following values: -10, -6, -3, -1, 3, and 4. The box plot shown is a graphic representation of this set. What is the sum of the values located at points A, B, and C?

In a box graph, the location of point B is the value of the median, which in this case is -2. We know that because with 6 terms in a set, the median is the average of the 2 center terms, which are -3 and -1. The location of point A is the median of the values less than the median. Since -10, -6, and -3 are the values less than the median (-2), this is -6. The location of point C is the median of the values more than the median. Since -1, 3, and 4 are the values more than the median, this is 3. Points A, B, and C are therefore -6, -2, and 3, and the sum of those values is -5.

7.

Drill Explanations | 159

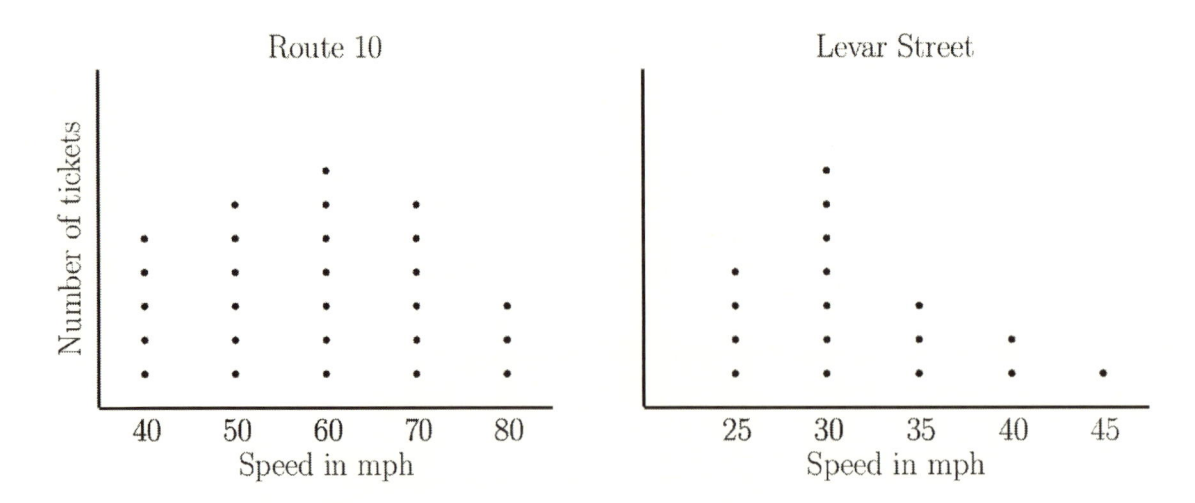

A record was kept of the speed of drivers as they passed 2 different points, 1 on Route 10 and the other on Levar Street. The dot plot shown is a record of the approximate speed of the drivers as they passed these points. Which of the following is true?

A) The standard discrepancy of the speed of the Route 10 drivers is greater than the standard discrepancy of the speed of the Levar Street drivers.
B) The standard discrepancy of the speed of the Route 10 drivers is less than the standard discrepancy of the speed of the Levar Street drivers.
C) The standard discrepancy of the speed of the Route 10 drivers is equal to the standard discrepancy of the speed of the Levar Street drivers.
D) There is not enough information to compare the standard discrepancy of the speed of the drivers on Route 10 and the speed of the drivers on Levar Street.

We can do this without trying to determine exact values. For Route 10, we see that most drivers have a speed of 60 mph. Furthermore, there are somewhat fewer drivers above 60 mph than below, but the difference isn't that great. This implies that the average is somewhere between 50 mph and 60 mph, though plenty of people are driving substantially slower or faster than that.

For Levar Street, the average is a little harder to approximate, but with so many drivers going 30 mph and none of the other speeds being very far from this, the average won't be too far from 30 mph, not that many drivers are going much slower or faster than that. This means that Levar's standard discrepancy is less than that of Route 10 where the speeds are "all over the place" compared to the "clumping" around 30 of Levar Street. That supports **choice A**.

8.

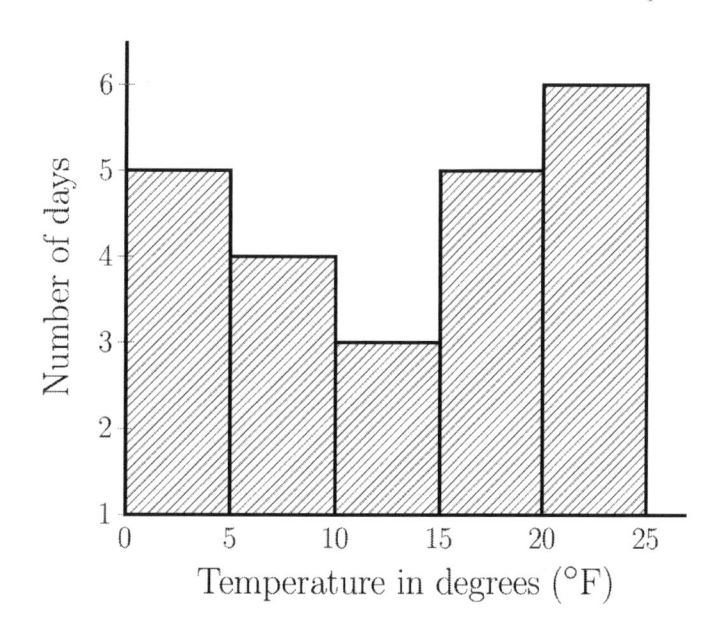

The average daily temperature at a certain location was recorded on 23 consecutive days. The histogram shown is a record of the coldest temperatures on those days. The 1st bar, for example, shows the number of days that the coldest temperature was at least 0 degrees and less than 5 degrees. Based on this graph, what was the coldest possible temperature for the day on which the median temperature was recorded, if this temperature has an associated margin of error of 5%?

With 23 terms (days), the median will be the 12th term if we lined up all the values in ascending order. The 1st 5 terms (the coldest 5 days) will all have temperatures of at least 0 degrees and less than 5 degrees. The next 4 terms will have temperatures of at least 5 degrees and less than 10 degrees. That's 9 terms so far. The next 3 terms, including the 12th term, will be in the next category, at least 10 degrees and less than 15 degrees. We don't know exactly what these terms are, but since the question asks about the coldest possible temperature for this day, we'll say that the 10th, 11th, and 12th terms each had a temperature of 10 degrees. There's no reason that couldn't be the case.

Taking into account the 5% margin of error, since we want the coldest possible temperature, we'll say that this temperature could actually be 5% less than 10 degrees, making it **9.5 degrees**.

9.

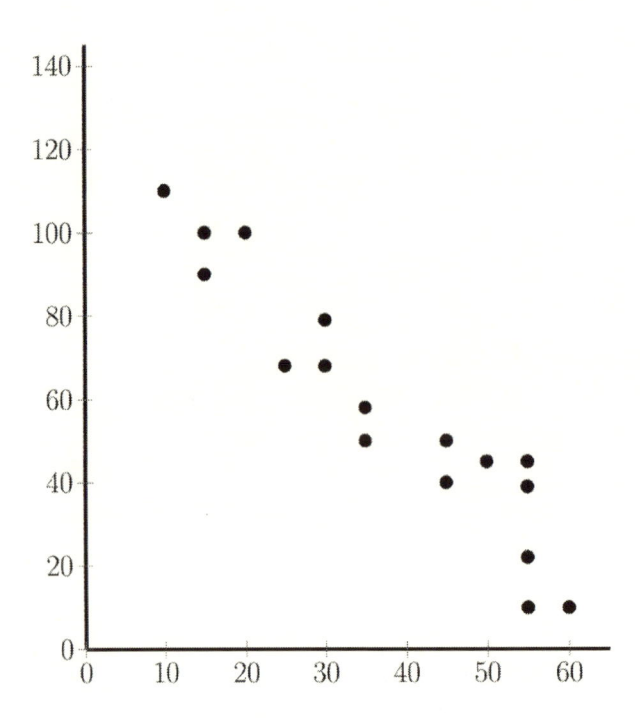

Which of the following equations is the most appropriate linear model for the data shown on the scatterplot?

A) $y = -x + 130$
B) $y = -\frac{11}{6}x + 60$
C) $y = \frac{11}{6}x + 130$
D) $y = -\frac{11}{6}x + 130$

It's clear that the most appropriate linear model, or simply the line of best fit, would have a negative slope, so we can eliminate choice C. The line, if extended, would have a y-intercept somewhat greater than 110, eliminating choice B. Now we'll want to approximate the slope to determine if the correct choice is -1, as in choice A, or -11/6 (which is quite close to -2) as in choice D. Let's be careful because the values on the 2 axes have different scales, making it harder to estimate the slope from a casual look at the data points.

If we look at the leftmost point, its location is about (**10, 110**). If we look at the rightmost point, its location is about (60, 18). Since the 2 slopes in the remaining choices are not all that close, let's round that location to (**60, 20**). For slope we'll divide the difference of the *y* coordinates of these 2 points by the difference of the *x* coordinates. That gives us 90 divided by -50, which is $-\frac{9}{5}$, which is close to -2, making **choice D** correct.

Day 13

Quiz 2

After reading this material, reviewing days 6–11, and then completing Quiz 2, sign and check the box for Day 13.

Note: We'll review Day 12 in a later quiz.

Today is another review day. Go back over any problematic areas, and then complete the quiz. You've had a lot of material to master in the last week, and it's crucial that you can access the associated skills without hesitation. We get to that stage through repetition and not getting stalled by the tougher stuff. Everything you need is on these pages.

We've recently looked at some of the content that shows up in the tougher questions—questions that *could* be tough if they're unfamiliar to you. Some of these questions can be done in less than 30 seconds if you can pull out the right tool from your expanding tool kit and apply yourself with maximum efficiency. Never rush. Aim for the confident smoothness that comes from review, re-review, and, if needed, re-re-review. Anything you sort of know should be attacked. We're looking for **mastery.** You can do it!

Quiz 2

1.

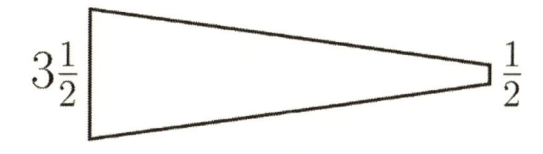

$3\frac{1}{2}$ $\frac{1}{2}$

The trapezoid shown has an area of 22. What is its height?

2.

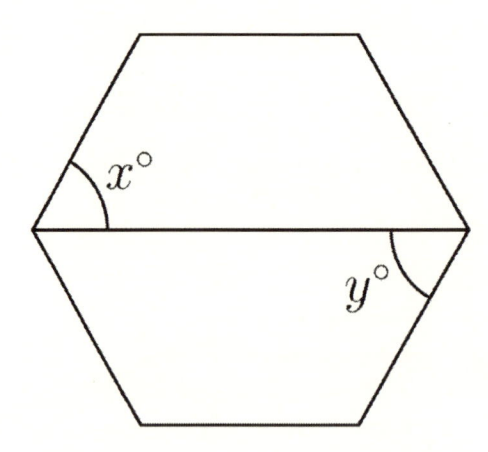

What is the sum of x and y in the bisected regular hexagon shown?

3. A right triangle has a leg that is ¼ inch and a hypotenuse that is ½ inch. What is the area of the triangle?

4. A right triangle has legs that are 5×10^7 centimeters and 12×10^7 centimeters. What is the length of the hypotenuse length expressed in scientific notation?

5.

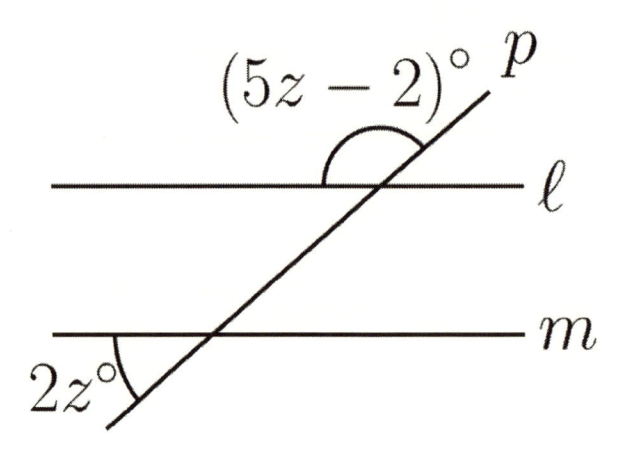

Lines *l* and *m* are parallel and intersected by line *p*. What is the value of *z*?

6.

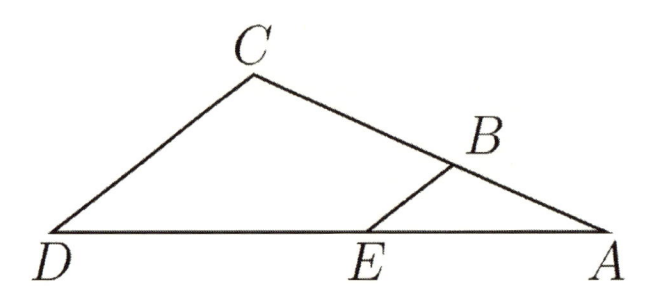

164 |Day 13

In the triangle shown, $\overline{BE}$ is parallel to $\overline{CD}$. If $\overline{AC} = 14$ and $\overline{AB} = 6$, $\overline{BE}$ is what fraction of $\overline{CD}$?

7. What is the distance in the coordinate plane between points $(-6,-1)$ and $(2,2)$?

8. What is the x-intercept of the line defined by the equation $y = -5x - 9$?

9. What is $\frac{\pi}{6}$ expressed in degrees?

10. What is the equation defining a circle with a vertex at $(0,-3)$ and a radius of 4?

11. What are the coordinates of the vertex of the parabola defined by the equation $y = x^2 - 8x + 9$?

12. What is the $ax^2 + bx + c$ form of the parabola defined by $-(x+2)^2 - 3$?

13. What are the coordinates of the midpoint of the x and y intercept of the line defined by $y = 2x - 9$?

14. What are the x-coordinates of the points where the line defined by $-x + 13$ and the parabola defined by $2x^2 - 3x + 1$ intersect?

15. The function f is defined by $f(x) = ax^2 - 4x + c$, in which a and c are constants. If $f(2) = 5$ and $f(-1) = 8$, what is the value of a?

16.

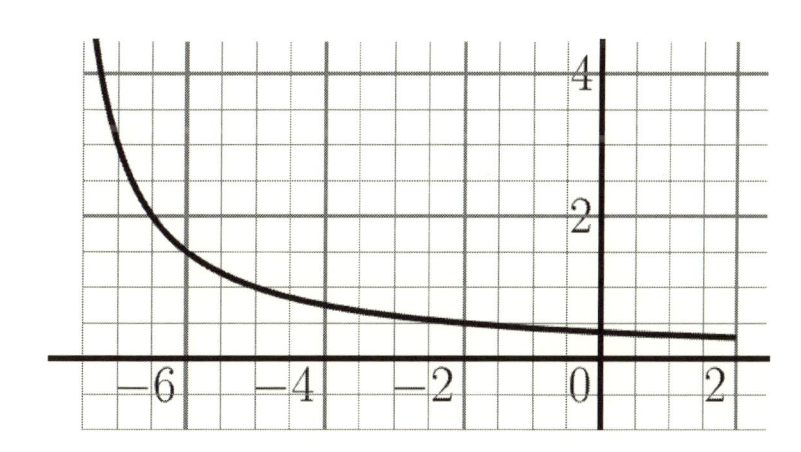

The rational function f is defined by the equation $f(x) = \frac{3}{x+r}$ where r is a constant. The partial graph of $y = f(x)$ is shown. Which of the following is the value of r?

A) 0.4
B) 1.75
C) 3
D) 8

Quiz 2 | 165

17.

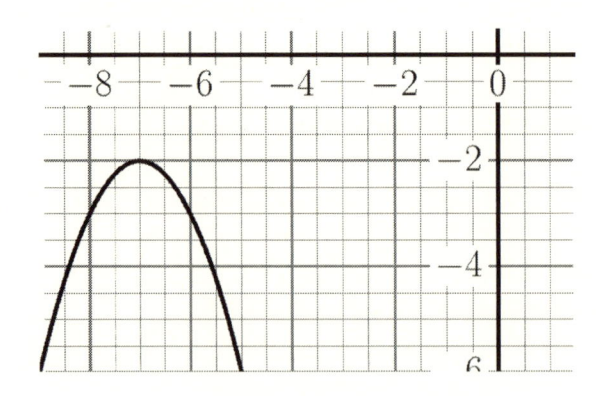

The graph of $y = h(x)$ is shown. If $h(x) = g(x+3)$, which of the following is the $g(x)$?

A) $-(x+4)^2 - 2$
B) $-(x-2)^2 + 4$
C) $(x+4)^2 - 2$
D) $(x+7)^2 - 2$

18. What is the circumference of a circle defined by the equation $(x+2)^2 + (y+3)^2 = 121$?

19.

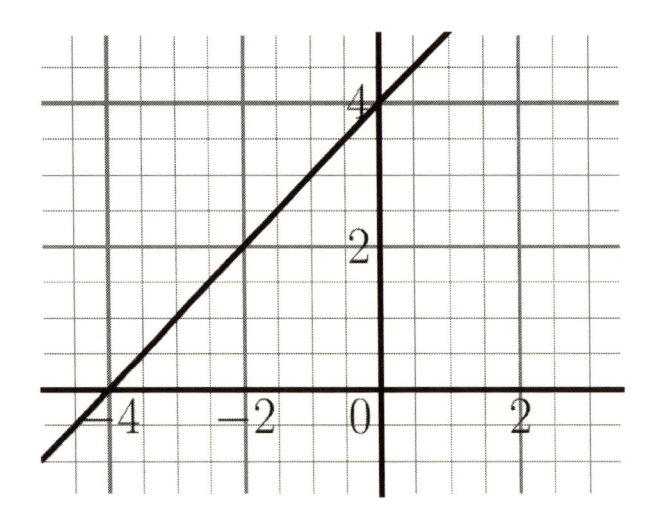

What are the coordinates of the point at which the line shown intersects the circle defined by $(x-3)^2 + (y+3)^2 = 50$?

A) -4
B) -2
C) 2
D) 4

20. A jewelry store uses the equation $f(m) = 0.5m - 6$, where $m \geq 14$ to determine how many repairs can be completed m minutes after the shop has opened. Which of the following is true?

A) Every 2 minutes after the store opens, an additional repair can be completed.
B) Every 30 seconds after the store opens, an additional repair can be completed.
C) The store can complete 6 repairs every 5 minutes.
D) The store can complete 6 repairs every 6 minutes.

21. Cecily spent 1 hour walking from her house to her friend's house b blocks away and then later walking back along the same route. On her way to the house, she walked at the rate of 1.5 blocks every minute. On her way back, she walked at a different rate. This situation can be represented by the equation $1.5(b) + b = 60$. Which of the following is true?

A) Cecily walked back from her friend's house at the rate of 1 block every minute.
B) Cecily walked back from her friend's house more quickly on average than she did walking to the house.
C) Cecily's friend lives 30 blocks from Cecily.
D) Cecily's friend lives 60 blocks from Cecily.

22. A college professor teaches 2 classes, and 1 of them has 3 more students than the other. The students in the larger class had a higher average score on a test given to both classes. If n represents the number of students in the larger class, S_1 represents the sum of the scores in the larger class, and S_2 represents the sums of the scores in the other class, which of the following represents the positive difference in the average score of the students in the 2 classes?

A) $\frac{S_1}{n} - \frac{S_2}{n+3}$
B) $\frac{S_1}{n+3} - \frac{S_2}{n}$
C) $\frac{S_1+3}{n} - \frac{S_2}{n}$
D) $\frac{S_1}{n} + \frac{S_2}{n+3}$

23. Javier deposited p dollars into a savings account that pays 3.5% interest annually. If no money is withdrawn or deposited after the initial deposit, which of the following represents the amount of money in the account after t years?

A) $p(0.035)^t$
B) $p(1.035)^t$
C) $t(1.035)^p$
D) $t(0.035)^p$

24. The function f is defined by the equation $f(x) = x^2 + 2x$. Which of the following shows the equivalent function with the minimum x and $f(x)$ values shown as constants or coefficients?

A) $f(x) = (x+2)^2 - 1$
B) $f(x) = (x+1)^2 - 1$
C) $f(x) = (x+1)^2$
D) $f(x) = (x-1)^2$

25.

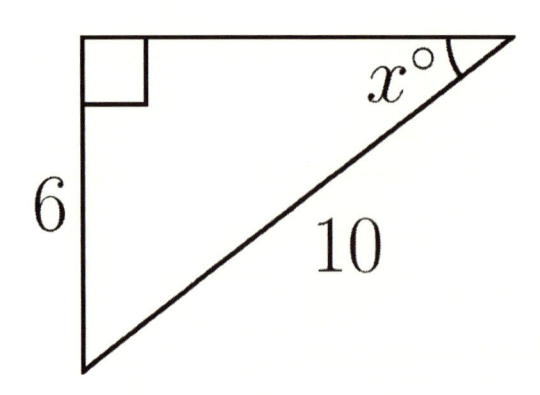

[triangle figure: right angle at top-left, vertical leg = 6, hypotenuse = 10, angle $x°$ at top-right]

What is the tan $x°$ in the triangle shown?

26. If $\frac{r}{s} = \frac{5}{4}$ and $r - s = 10$, what is $\sin(r+s)°$?

27. Triangle ABC has angles of measurements a, b, and c degrees. If $\sin a° = \cos b°$, which of the following must be true?

A) $a = b$
B) $a \neq b$
C) ABC is an equilateral triangle.
D) ABC is a right triangle.

Day 14

Rest, Review, Quiz 2 Explanations

Quiz 2 Explanations

1.

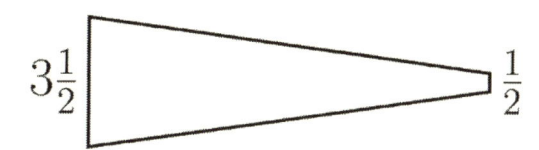

$3\frac{1}{2}$ $\frac{1}{2}$

The trapezoid shown has an area of 22. What is its height?

The area of a trapezoid is the product of "the average of the bases" and the height. Here, the average of the bases, the 2 given sides, is 2. Since the area is 22, the height must be **11**.

2.

$x°$

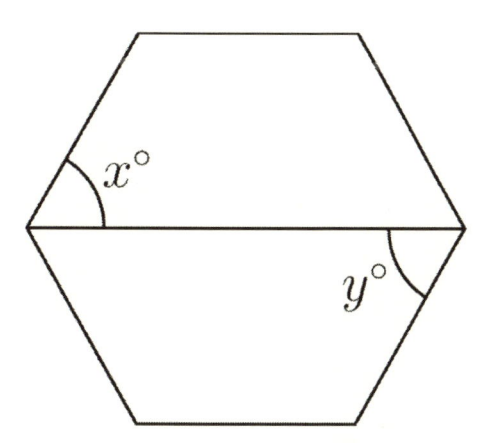

$y°$

What is the sum in degrees of x and y in the bisected regular hexagon shown?

We get the sum of the angles of a polygon by using $180(n-2)$ where n is the number of sides. This has 6 sides, so that's a sum of 720 degrees. Because this is a regular hexagon, all the sides are the same, and all the angles are the same. That means each angle is 1/6 of 720, which is 120 degrees. The 2 angles shown are each half of this, so together they sum to **120**.

3. A right triangle has a leg that is ¼ inch and a hypotenuse that is ½ inch. What is the area of the triangle in square inches?

Always ask yourself if you have a special right triangle. Since the hypotenuse is twice one of the legs, we know that this is the 30/60, and the longer leg is root 3 times the shorter leg, or $\frac{\sqrt{3}}{4}$. The area of a triangle is half the product of the base and height. We can get that by taking half of the product of the legs: $\frac{1}{4} \times \frac{\sqrt{3}}{4} = \frac{\sqrt{3}}{16}$, and half of that is $\frac{\sqrt{3}}{32}$.

4. A right triangle has legs that are 5×10^7 centimeters and 12×10^7 centimeters. What is the length of the hypotenuse length expressed in scientific notation?

This is a special right triangle: the 5, 12, 13. Remember to look for them! That means the hypotenuse is simply 13×10^7

5.

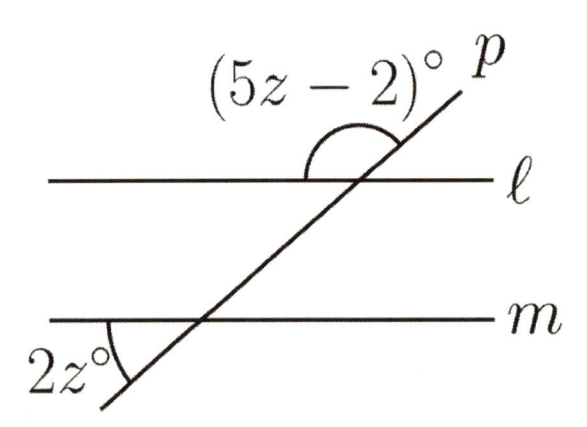

Lines *l* and *m* are parallel and intersected by line *p*. What is the value of *z*?

Since these are parallel lines intersected by a 3rd line, 4 angles have been created: 4 "little ones" and 4 "big ones." The sum of a little and a big is 180. Therefore, $(5z - 2) + 2z = 180$. That simplifies to $7z = 182$, and $z = 26$.

6.

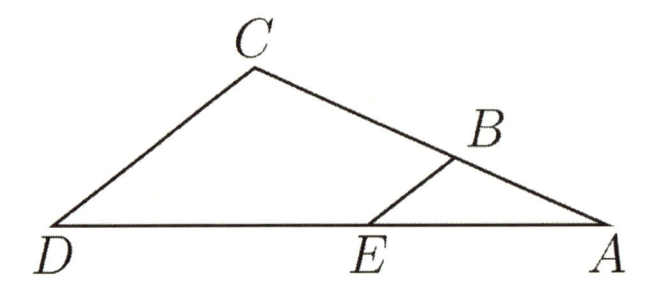

In the triangle shown, $\overline{BE}$ is parallel to $\overline{CD}$. If $\overline{AC} = 14$, and $\overline{AB} = 6$, $\overline{BE}$ is what fraction of $\overline{CD}$?

170 |Day 14

Knowing that the 2 segments are parallel, we know that the 2 triangles are similar. This means that their corresponding sides have the same proportions. We know that $\overline{AB}$ is $\frac{6}{14}$ of $\overline{AC}$, which simplifies to $\frac{3}{7}$. This means that $\overline{BE}$ is also $\frac{3}{7}$ of $\overline{CD}$.

7. What is the distance in the coordinate plane between points $(-6, -1)$ and $(2, 2)$?

If you know the distance formula, that's great. Otherwise, you can triangulate on the coordinate plane. Once you make the 2 given points the endpoints of a triangle's hypotenuse, you'll see that the legs are 3 and 8. This is **not** a special triangle, so using the Pythagorean theorem, we see that the hypotenuse (that's what we want) is the square root of the sum of 9 and 64. That gives us the irrational number $\sqrt{73}$.

8. What is the x-intercept of the line defined by the equation $y = -5x - 9$?

The x-intercept occurs where the y-coordinate is 0. That gives us $0 = -5x - 9$, which simplifies to $x = -\frac{9}{5}$. Note: -1.8 is also acceptable.

9. What is $\frac{\pi}{6}$ expressed in degrees?

Keep in mind that 180 degrees equals π radians. Therefore, we want $\frac{1}{6}$ of 180, which is **30**.

10. What is the equation defining a circle with a vertex at $(0, -3)$ and a radius of 4?

$x^2 + (y + 3)^2 = 16$. Don't know this equation yet? It's in Day 8.

11. What are the coordinates of the vertex of the parabola defined by the equation $y = x^2 - 8x + 9$?

You could use the OSC to graph this, and then you'll easily see the vertex. Or you could complete the square (from Day 4), giving you the equation $(x - 4)^2 - 7$. Now we know the vertex is at $(4, -7)$.

12. What is the $ax^2 + bx + c$ form of the parabola defined by $-(x + 2)^2 - 3$?

If you expand the equation, first squaring $(x + 2)$, then applying the negative sign, and then subtracting 3, you'll get $-x^2 - 4x - 7$.

13. What are the coordinates of the midpoint of the x and y intercepts of the line defined by $y = 2x - 9$?

The x-intercept occurs where the y-coordinate is 0, and the y-intercept occurs where the x-coordinate is 0.

Rest, Review, Quiz 2 Explanations | 171

So we have $0 = 2x - 9$, which simplifies to $x = \frac{9}{2}$, and we have $y = 2(0) - 9$, which simplifies to $y = -9$. The intercepts are therefore at $(\frac{9}{2}, 0)$ and $(0, -9)$. The average of the x-coordinates of these points is $\frac{9}{4}$, and the average of their y-coordinates is $-\frac{9}{2}$. Thus, the midpoint is at $(\frac{9}{4}, -\frac{9}{2})$.

14. What are the x-coordinates of the points where the line defined by $y = -x + 13$ and the parabola defined by $y = 2x^2 - 3x + 1$ intersect?

You can use the OSC, or you can set $-x + 13$ equal to $2x^2 - 3x + 1$. If we solve it this way, the equation simplifies to $2x^2 - 2x - 12 = 0$. After dividing each side by 2, we get $x^2 - x - 6 = 0$, which factors into $(x - 3)(x + 2) = 0$, which means that the x-coordinates at the intersection of the line and parabola are 3 and -2.

15. The function f is defined by $f(x) = ax^2 - 4x + c$, in which a and c are constants. If $f(2) = 5$ and $f(-1) = 8$, what is the value of a?

We can't graph the function equation without knowing the value of a, so instead, we'll plug the values into the function equation, $5 = a(2)^2 - 4(2) + c$ as well as $8 = a(-1)^2 - 4(-1) + c$. The 1st equation simplifies to $13 = 4a + c$, and the 2nd equation simplifies to $4 = a + c$. If we subtract the second equation from the 1st, we get $9 = 3a$, which means $a = 3$.

16.

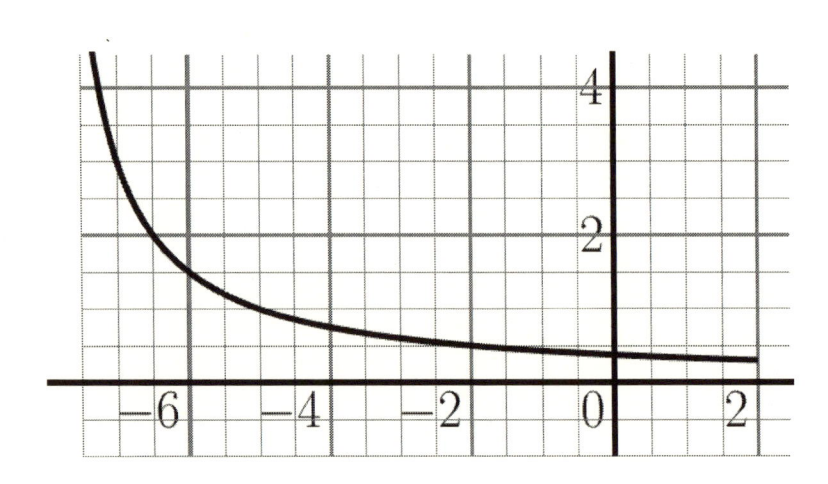

The rational function f is defined by the equation $f(x) = \frac{3}{x+r}$ where r is a constant. The partial graph of $y = f(x)$ is shown. Which of the following is the value of r?

A) 0.4
B) 1.75
C) 3
D) 8

We can see that the point $(-6, 1.5)$ appears to lie on the graph. (So does, for example, $(-2, 0.5)$, but we don't need to use 2 points.) We can now plug the coordinates into the equation to get $1.5 = \frac{3}{-6+r}$. This simplifies to $-9 + 1.5r = 3$, which further simplifies to $r = 8$.

17.

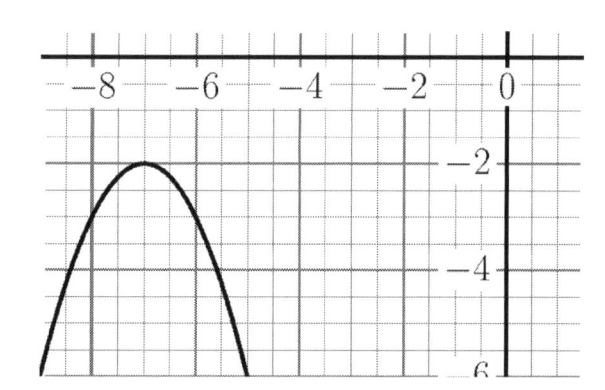

The graph of $y = h(x)$ is shown. If $h(x) = g(x + 3)$, which of the following is the $g(x)$?

A) $-(x + 4)^2 - 2$
B) $-(x - 2)^2 + 4$
C) $(x + 4)^2 - 2$
D) $(x + 7)^2 - 2$

The graph of the function h appears to have a vertex at $(-7, -2)$ and opens downward. It can therefore be defined by the equation $h(x) = -(x + 7)^2 - 2$. (We know there is no coefficient before $(x + 7)^2$ other than negative 1 because there is none in any of the choices.) Since $h(x) = g(x + 3)$, the function h has x values that are 3 more than that of function g. Put another way, the x values of g are 3 **less**. Therefore, $g(x) = -(x + 4)^2 - 2$, which is **choice A**.

You could also solve this one efficiently by realizing that we still need a downward-opening parabola (eliminating choices C and D) and the y-coordinate of the vertex will still be -2, which eliminates choice B.

18. What is the circumference of a circle defined by the equation $(x + 2)^2 + (y + 3)^2 = 121$?

If you know the equation that defines a circle, then you know the radius of this circle is 11. That means the diameter is 22 and the circumference is 22π.

19.

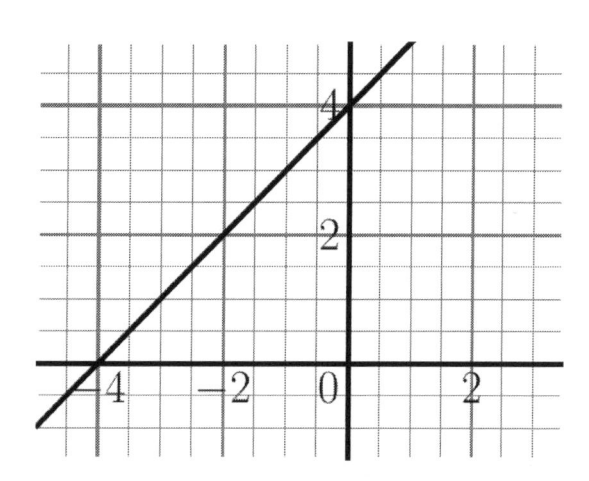

Rest, Review, Quiz 2 Explanations | 173

What is the y-coordinate of the point at which the line shown intersects the circle defined by $(x-3)^2 + (y+3)^2 = 50$?

A) -4
B) -2
C) 2
D) 4

We can see that the line has the equation $y = x + 4$. Since we want the intersection point with the circle, we can substitute $x + 4$ for y in the circle equation, giving us $(x-3)^2 + (x+7)^2 = 50$. Expanding this equation, combining like terms, and then setting it equal to 0 give us $2x^2 + 8x + 8 = 0$. We can now divide both sides by 2 and then factor, giving us $(x+2)(x+2) = 0$. That means the only solution (the only intersection point) occurs when $x = -2$. If we plug this into our line equation, we see that $y = 2$.

Can we use the OSC here? Yes. Once we recognize that the line's equation is $y = x + 4$, we can graph it and then graph the circle. The intersection point is then readable. Can we know for sure just looking at the graph that the line really is $y = x + 4$ and not, for example, $y = x + 4.001$? No, **but** our answer choices are not that close to each other, and the test isn't deceptive in that way.

20. A jewelry store uses the equation $f(m) = 0.5m - 6$, where $m \geq 14$ to determine how many repairs can be completed m minutes after the shop has opened. Which of the following is true?

A) Every 2 minutes after the store opens, an additional repair can be completed.
B) Every 30 seconds after the store opens, an additional repair can be completed.
C) The store can complete 6 repairs every 5 minutes.
D) The store can complete 6 repairs every 6 minutes.

Let's try out some values. If it's 14 minutes after the store opens, then 1 repair is completed. At 15 minutes, only 1.5 repairs are done. But at 16 minutes, 2 repairs are completed. At 18 minutes, 3 repairs are done. That's 1 additional repair every 2 minutes, leading us to **Choice A**.

21. Cecily spent 1 hour walking from her house to her friend's house b blocks away and then later walking back along the same route. On her way to the house, she walked at the rate of 1.5 blocks every minute. On her way back, she walked at a different rate. This situation can be represented by the equation $1.5(b) + b = 60$. Which of the following is true?

A) Cecily walked back from her friend's house at the rate of 1 block every minute.
B) Cecily walked back from her friend's house more quickly on average than she did walking to the house.
C) Cecily's friend lives 30 blocks from Cecily.
D) Cecily's friend lives 60 blocks from Cecily.

The 1st term—$1.5b$—represents her rate (1.5 blocks per minute), multiplied by the number of blocks (b). The 2nd term—just b—represents the same thing, so the rate is 1 block per minute. The 1, of course, is implied, not written. That leads us to **choice A**.

174 |Day 14

22. A college professor teaches 2 classes, and 1 of them has 3 more students than the other. The students in the larger class had a higher average score on a test given to both classes. If n represents the number of students in the smaller class, S_1 represents the sum of the scores in the larger class, and S_2 represents the sums of the scores in the other class, which of the following represents the positive difference in the average score of the students in the 2 classes?

A) $\frac{S_1}{n} - \frac{S_2}{n+3}$
B) $\frac{S_1}{n+3} - \frac{S_2}{n}$
C) $\frac{S_1 + 3}{n} - \frac{S_2}{n}$
D) $\frac{S_1}{n} + \frac{S_2}{n+3}$

Since average is equal to sum divided by number of terms, and since the class with 3 more students had the higher average score, **choice B** is correct.

23. Javier deposited p dollars into a savings account that pays 3.5% interest annually. If no money is withdrawn or deposited after the initial deposit, which of the following represents the amount of money in the account after t years?

A) $p(0.035)^t$
B) $p(1.035)^t$
C) $t(1.035)^p$
D) $t(0.035)^p$

We need to multiply by 1.035 to get the amount deposited **plus** the interest, so that eliminates choices A and D. We want to raise this figure by the number of years and then multiply that by the initial deposit, leading us to **choice B**.

24. The function f is defined by the equation $f(x) = x^2 + 2x$. Which of the following shows the equivalent function with the minimum x and $f(x)$ values shown as constants or coefficients?

A) $f(x) = (x+2)^2 - 1$
B) $f(x) = (x+1)^2 - 1$
C) $f(x) = (x+1)^2$
D) $f(x) = x(x+2)$

You could graph this with the OSC and then read the vertex $(-1, -1)$ from the graph. You can then plug the coordinates into the vertex form of the equation, giving us **choice B**. You can also efficiently complete the square, leading to the same equation.

25.

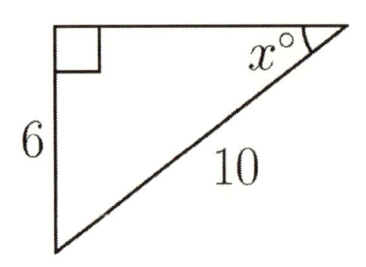

What is the tan $x°$ in the triangle shown?

This is a special right triangle, the 3:4:5. If we double each of these values, we get 6, 8, and 10. The missing side is therefore 8, making the tangent of x equal to 6 divided by 8 or 3/4. Note: 0.75 is also acceptable.

26. If $\frac{r}{s} = \frac{5}{4}$ and $r - s = 10$, what is $\sin(r + s)°$?

These simultaneous equations can be solved to get $r = 50$ and $s = 40$. The question, therefore, is what is the sine of 90 degrees? If you haven't already, you should commit to your memory the sine of 90. It's **1**, the largest possible sine of any angle.

27. Triangle ABC has angles of measurements *a, b,* and *c* degrees. If $\sin a° = \cos b°$, which of the following must be true?

A) $a = b$
B) $a \neq b$
C) ABC is an equilateral triangle.
D) ABC is a right triangle.

If the sine of an angle is equal to the cosine of another angle, the angles sum to 90 degrees. That means the other angle of the triangle must be 90 degrees (since the sum is always 180). We now know that we have a right triangle, as **choice D** says.

Day 15

Break

---

*After doing something—anything **besides** studying for the SAT—get outside. Exercise. Complete a school assignment. Take a nap. Then you can sign and check the box for Day 15.*

My mission is to make SAT Math easier for everyone, and the only way for me to accomplish that mission is by reaching…well...everyone.

This is where you come in. Most people do, in fact, judge a book by its cover (and its reviews). So here's my ask on behalf of a struggling student you've never met:

Please help that student by leaving this book a review.

By taking just a moment to leave an honest review on Amazon, you'll be helping future test-takers find the guidance they need. Your insights and experiences can make a big difference in someone else's study plan.

Your gift costs no money and less than 60 seconds to make real, but can change a fellow student's life forever.

Simply scan the QR code below to leave your review:

Thank you for your support and for being a part of this learning community!

Day 16

Advanced Algebra – Intricate Equations, No Solutions, Infinite Solutions

After reading this material and completing the drill at the end, sign and check the box for Day 16.

By now, you've been exposed to nearly all the subject matter you'll encounter on the day of your exam. However, the SAT folks do a good job of varying these concepts to the point that you might not recognize the concept. Sometimes the solution is to simply plow through some fairly intricate algebraic equations. Sometimes the solution is to realize that you can avoid most of that with a shortcut. And once in a while, we see some questions that really do require a bit of imaginative thinking. We'll cover all these possibilities today, first focusing on intricate equations.

Intricate Equations

$$2x - y^2 = \frac{4+z}{3}$$

Which of the following is equivalent to the equation above?

A) $z = y\sqrt{6x - z - 4}$
B) $x = \sqrt{y - z - 10}$
C) $y - z = \pm\sqrt{6x - 4}$
D) $y = \pm\sqrt{2x - \frac{z}{3} - \frac{4}{3}}$

Usually, it's a good idea to get rid of any fractions in an equation, which we can do here by multiplying each side by 3. We also know that this is a good idea because the answer choices don't have fractions. This gives us $6x - 3y^2 = 4 + z$. From here on, we'll use **this** form of the equation they gave us. What now? Let's be strategic and **consult the choices** for clues. Choice A isolates z, but if we isolate z by subtracting 4 from each side of our equation, we won't get anything that looks much like choice A.

Your intuition might already be guiding you to another choice to try or eliminate. Choice B isolates x. We can also do that (in our heads, perhaps) by taking our equation and thinking about what it would look like if we got *x* by itself. We would add $3y^2$ to each side and then divide by 6. That doesn't look much like choice B, does it?

Now let's isolate *y*, as in choice D. If it's not right, we'll choose C. To isolate *y*, we'll add $3y^2$ to each side **and** at the same time subtract $4 + z$ from each side. (We do it this way because we'll end up with a positive $3y^2$ on each side, not a negative one.) This gives us $6x - 4 - z = 3y^2$. Now to further isolate *y*, we'll divide each side by 3 **and** at the same time take the square root of each side.

And we're done. If you carefully divide each term by 3 and then take the square root of it all, that gives us **choice D**. Notice that we needed the plus/minus sign or the equation would be incomplete. This is like saying that if you know $x^2 = 36$, there are 2 solutions: 6 and negative 6.

What about using the **OSC** for that one?

It'll work if you're careful. You could enter the original equation and then see which choice gives you the same graph. However, the original equation has that 3rd unknown—*z*. If you enter it, the OSC will give you a slider option, and that can be a little confusing. What you **can** do is just choose a value for *z* and be consistent with it. For example, you could let $z = 1$. Once you get a graph, you would then enter the answer choices using 1 for *z* in each case. There's another issue, however. There's no simple way to enter the $\pm$ sign on the calculator. What you could do, though, is just enter it with a positive value (which for choice D will give you half the graph exactly, so you already know it's probably correct) and then copy what you've entered into the next entry field and add the negative sign where it belongs. That now matches completely. Of course, if you do this for every choice, starting with A, it could take a while.

Now we'll look at a question that we *do* need to solve.

What are the solutions to the equation $x^2 + 2x = 4$?

A) $\sqrt{5} \pm 1$
B) $-1 \pm \sqrt{5}$
C) $-1 \pm 2\sqrt{5}$
D) ± 2

Let's first see what happens when we enter this equation into the OSC. You'll see that you get a decimal that is an approximation of the solution and doesn't look like any of the choices. So what else can we do? We can set it equal to 0, but $x^2 + 2x - 4 = 0$ doesn't easily factor. What about completing the square? (If you're shaky on that, go back to Day 4.) Let's try it.

Completing the square gives us $(x + 1)^2 - 5 = 0$. We want to isolate *x*, so we add 5 to both sides, and then taking the roots of each side, we get $\pm(x + 1) = \sqrt{5}$. Like the last question, notice that we need the +/- sign when taking the root of a squared term. That means $x + 1 = \sqrt{5}$ and $-x - 1 = \sqrt{5}$. If we isolate *x*, we get **choice B**, though this is perhaps not immediately obvious.

Advanced Algebra – Intricate Equations, No Solutions, Infinite Solutions | 179

What about using the good ol' quadratic formula? It always works. If you plug in the values and simplify a bit, you get $\frac{-2 \pm \sqrt{20}}{2}$, which is the same as $\frac{-2 \pm 2\sqrt{5}}{2}$. We're not quite there. We can further simplify once we divide both of the terms on top by 2, which is, again, choice B.

Both of those methods are somewhat time-consuming, although they are not so bad if you can efficiently complete the square or use the quadratic formula. (And you **should** be able to do both of those without much stress.) But let's think a little more about the **OSC**.

We said that entering the equation gives us decimal approximations of 1.23606 and −3.23606. These numbers continue from there because they're irrational numbers. No, they don't look like choice B at a glance, but let's do some estimation.

Choice A is the square root of 5 plus 1. The square root of 5 is a little more than 2 (the square root of 4). So this choice is more than 3, which is definitely not one of the decimal figures we got with the OSC. We can also easily eliminate choice D since, well, that's not the answer! You could guess from here if you need to because of time, but let's take a closer look at choice C. If we consider only $-1 + 2\sqrt{5}$ for now, we have a number that is 1 less than a number that is greater than 4. (Remember that the square root of 5 is more than 2.) This results in a number that is greater than 3, so that's wrong. Choice B is looking good. $-1 + \sqrt{5}$ is a little more than 1, while $-1 - \sqrt{5}$ is a little less than −3. Those values are close to 1.23606 and −3.23606. So the OSC, together with some approximating of our own, is actually not that bad here.

Let's see what we can do with this next one.

In a research lab, a lizard completed a maze that was 30 feet in length. The next day, its speed for completing the maze was 1.5 times faster than on the previous day, allowing it to complete the maze in 2 seconds less time than on the previous day. What was the average speed on the 1st day?

If you were given answer choices, you could try them out. Remember, that's a useful strategy when the choices are known values. So once you tried the right answer—which is **5 feet per second**—you would just enter that into the word problem. The question is about the speed on the 1st day, so you'd say that going 30 feet at a rate of 5 feet per second gives us a time of **6 seconds**. The next day it traveled 7.5 feet per second (that's 5 x 1.5). Dividing 30 by the faster rate of 7.5 feet per second gives us **4 seconds**. This is, indeed, 2 seconds less time, so we know it's correct.

Of course, you don't need to try the answers out to get the question right. That can be time-consuming. Let's see about using our algebraic skills.

If r is the 1st day rate (that's ultimately what we're looking for) and t is the 1st day time, we can say that $\frac{30}{r} = t$.

We're simply applying our rate formula, which says that rate x time equals distance (or distance divided by rate equals time). We need a 2nd equation and that would be $\frac{30}{1.5r} = t - 2$. In other words, going 1.5 times faster gets the lizard there in 2 fewer seconds.

180 |Day 16

From there you could use substitution to solve for these 2 equations. Can the OSC help us out, though? Well, it won't want to know about r and t, but if you use x and y instead, you will see 2 intersecting points. Why 2? Well, you end up with a quadratic, but in our real-world situation with the lizard and the maze, we can't have a negative speed. Assuming that you used x for r, then the x-coordinate at the intersecting point is, of course, **5 feet per second.**

Here's another question that asks us **not** to solve but to **represent** a real-world situation with algebra.

At a factory, Machine X and Machine Y run for h hours each day, during which they produce parts for $\frac{5}{6}$ of that time. Machine X produces n parts in an hour, and Machine Y produces $n + 50$ parts in an hour. Which of the following expressions represents the total number of parts the 2 machines produce each day?

A) $\frac{5h(n+25)}{3}$

B) $\frac{h(5n+50)}{3}$

C) $\frac{5hn+125)}{3}$

D) $\frac{5h(n+1)}{6}$

A few approaches are possible here, including picking your own values. Just make sure to pick values that make the math simple. As always, you'll want to avoid choosing 1 or choosing the same number for 2 different unknowns. Here we could make **h, the number of hours, equal to 6** since it's easy to take 5/6 of that. And we could say that **n is 100**. If we plug those numbers into the question, the answer is 1,250 parts. If you plug those same numbers in for h and n in the choices, only choice A gives you 1,250. We're done.

Of course, you might not try choice A first, but let's approach this strategically. This is a later question, so the answer shouldn't be too obvious. That makes choice B not so great because it contains 50, which we find in the question. You might try choice D since it's simple, and you'll quickly see that it doesn't work out to be 1,250. Then you're quickly down to 2 choices. If you were to try choice C next, once you saw that it didn't give you 1,250, you could then simply choose A without having to test it. (Of course, if you do have time, it's not a bad idea.)

As always, this can be done algebraically. Each machine runs for $\frac{5}{6}h$ hours each day. Since Machine X produces n parts each hour, that's $\frac{5}{6}hn$ parts in total each day. Likewise, Machine Y produces $\frac{5}{6}h(n+50)$ each day. Together, that's $\frac{5}{6}hn + \frac{5}{6}h(n+50)$ each day. (You could now graph this, using x and y, and then see what gives you the same graph as 1 of the choices, but let's carry on with the algebra.)

Distributing the 2nd term, we get $\frac{5}{6}hn + \frac{5}{6}hn + \frac{250h}{6}$. We can add the first 2 terms and simplify the last one to get $\frac{5hn}{3} + \frac{125h}{3}$. We're almost there. We can add these, giving us $\frac{5hn+125h}{3}$, and finally factor the $5h$ from the 2 terms in the numerator, giving us, once again, choice A. Remember, factoring is a step that many students forget. But not you!

Advanced Algebra – Intricate Equations, No Solutions, Infinite Solutions | 181

No Solution

The test often asks tricky algebraic questions involving equations with **no** solution. These don't necessarily look intricate.

$$3x - 1 = bx + 24$$

In the equation above, b is a constant. What value of b would result in an equation with no solution?

This would probably be a question without answers to choose from. It can be done in no time if you realize that this equation would not hold true if $3x$ were to equal bx. In other words, if $b = 3$, then the equation would read $3x - 1 = 3x + 24$. This is impossible if you think about it (or if you try to solve it.)

Here's a somewhat more complex example of this same idea.

$$5 - abk = 17 - 3ab$$

In the equation above, k is a constant. What value of k would result in an equation with no solution?

Do you see how similar this is? The coefficient of ab is 3 on the right side of the equation. If it were **3** on the other side, the equation can't be solved. It would read $5 - ab3 = 17 - 3ab$, or simply $5 - 3ab = 17 - 3ab$, and therefore there is no solution.

Here's another way this idea can be tested.

Line l is defined by the equation $y = 4x + 1$. If line m is defined by the equation $rx = -8.5 + 0.5y$ such that r is a constant, for what value of r would the lines not intersect?

There is only one way that 2 distinct lines would not intersect, and that's because they're parallel. What would make them parallel is if they had the same slope. We can easily determine that the slope of line l is 4. Now we need to get the other equation into the same form.
Let's isolate y, which we can begin to do by doubling every term in this equation: $2rx = -17 + y$. If we isolate y, we get $y = 2rx + 17$. So if r is equal to **2**, the equation becomes $y = 4x + 17$, giving it the same slope as line l, and therefore they would not intersect. Put another way, given these 2 equations, there is no solution.

Infinite Solutions

Besides seeing problems with **no** solution, we also see problems on the test that involve having **infinite** solutions.

In the equation $3x + b + 5 = 3x$, for what value of the constant b does the equation have infinite solutions?

There are a few ways to approach this. An infinite solution means that x can be any number. If $b + 5 = 0$, then the equation would read $3x = 3x$, and in that case, x can be any number. Therefore, if b is equal to -5, the equation has infinite solutions.

Day 16

Drill

Complete all the questions below.

1.

$$\frac{6a^2 - b}{3} = a + 4c$$

Which of the following is equivalent to the equation above?

A) $\frac{1}{2}a^2 - \frac{1}{4}a - \frac{1}{12}b = c$
B) $a^2 - \frac{1}{2}a + \frac{1}{2}b = c$
C) $a + \frac{1}{2}a + \frac{1}{12}b = c$
D) $a^3 - \frac{1}{2}a + \frac{1}{12}b = c$

2. What are the solutions to the equation $x^2 - 8 = -4x$?

A) $-3 \pm 2\sqrt{3}$
B) $-2 \pm 3\sqrt{2}$
C) $-2 \pm 2\sqrt{3}$
D) $2 \pm 2\sqrt{3}$

3. A medical laboratory uses a machine to clean metal supplies. The machine once performed 3.6 cleanings every hour. Because of an increase in the number of supplies being cleaned, the machine now performs only 2.8 cleanings every hour. If it now takes the machine 4 hours longer to perform c cleanings, what is the value of c?

A) 8.0
B) 14.2
C) 50.4
D) 54.4

4. During an experiment, researchers determined that the equation $n = -\frac{3}{5}h + \frac{53}{5}$ can be used to model the number of bacteria in thousands (n) remaining in a Petri dish h hours after the start of the

experiment. If the experiment began at 12:00 p.m., how many more bacteria remained in the dish the next day at 4:00 a.m. than remained in the dish earlier at 2:20 a.m.?

We have a linear equation with 2 unknowns but are then given a value to use for 1 of the unknowns. Using 16 for h, we get $n = -\frac{3}{5}(16) + \frac{53}{5}$. This simplifies to $n = 1$. Let's not forget that n represents the number of bacteria *in thousands*, so the answer is 1,000.

5. The average selling price of a house in a neighborhood of 30 houses is d dollars. The average selling price of a house in a nearby neighborhood of 25 houses is $12,500 less than that. Which of the following represents the average selling price of all the houses?

A) $55d - 12,500$
B) $\frac{d}{30} + \frac{d-12,500}{25}$
C) $\frac{11d - 62500}{7}$
D) $55d - 312,500$

6. The equation $1 + st = s + t$ has infinite solutions if which of the following is true?

A) $s = -t$
B) $st = 2$
C) $t = 1$
D) $s = t$

7.
$$3y = \frac{3x}{2} - 9$$
$$2y = x$$

The equations above define 2 lines. How many intersecting points do they have?

A) 0
B) Exactly 1
C) Exactly 2
D) Infinitely many

Day 16

Drill Explanations

1.
$$\frac{6a^2 - b}{3} = a + 4c$$

Which of the following is equivalent to the equation above?

A) $\frac{1}{2}a^2 - \frac{1}{4}a - \frac{1}{12}b = c$
B) $a^2 - \frac{1}{2}a + \frac{1}{2}b = c$
C) $a + \frac{1}{2}a + \frac{1}{12}b = c$
D) $a^3 - \frac{1}{2}a + \frac{1}{12}b = c$

Let's first "flatten" the equation by multiplying each side by 3. That gives us $6a^2 - b = 3a + 12c$. The choices all have c isolated, so let's subtract $3a$ from each side and then divide by 12. That gives us $\frac{6a^2 - b - 3a}{12} = c$. Now let's split this into $\frac{6a^2}{12} - \frac{b}{12} - \frac{3a}{12}$. Doing some cancellations, we get $\frac{a^2}{2} - \frac{b}{12} - \frac{a}{4}$. With a little rearranging, this is the same as **choice A**.

2. What are the solutions to the equation $x^2 - 8 = -4x$?

A) $-3 \pm 2\sqrt{3}$
B) $-2 \pm 3\sqrt{2}$
C) $-2 \pm 2\sqrt{3}$
D) $2 \pm 2\sqrt{3}$

The choices suggest the quadratic formula, so we'll first add $4x$ to each side to get it equal to 0. This gives us $x^2 - 8 + 4x = 0$, and that means $a = 1, b = 4, c = -8$. (Remember, c is always the constant.) If we plug this into the quadratic formula, we get $x = \frac{-4 \pm \sqrt{4^2 - 4(1)(-8)}}{2}$. That works out to be $x = \frac{-4 \pm \sqrt{48}}{2}$. You could use the OSC to get irrational numbers and then see if you can figure out which choice matches, but let's continue. Root 48 can be factored into root 16 times root 3, or simply $4\sqrt{3}$. We now have $x = \frac{-4 \pm 4\sqrt{3}}{2}$. Once we divide each of the 2 terms in the numerator by 2, we get **choice C**.

3. A medical laboratory uses a machine to clean metal supplies. The machine once performed 3.6 cleanings every hour. Because of an increase in the number of supplies being cleaned, the machine now

performs only 2.8 cleanings every hour. If it now takes the machine 4 hours longer to perform c cleanings, what is the value of c?

A) 8.0
B) 14.2
C) 50.4
D) 54.4

Since the choices are known values (not x, y, etc.), we could try them. Looking at the choices, you might think this will take a while. However, let's be strategic. We know that the difference in time is exactly 4 hours. To get the time, we'll divide the number of cleanings by the hourly rate. (This is like saying divide the *distance* by the rate.)

Knowing the test, there's a good chance that dividing the correct number of cleanings by the rate will give us integer values. It should be clear without too much effort that choice A will not give us integers when we divide them by either rate. You could try the others without much effort. Once we get to choice C, you'll see that dividing 50.4 by 3.6 gives us 14 hours—an integer—and when we divide 50.4 by 2.8, we get 18 hours. This is a difference of 4 hours, so we're done. The answer is **choice C**.

Otherwise, we could set this up algebraically as $\frac{c}{2.8} - \frac{c}{3.6} = 4$. You could deal with this by finding a common denominator and going from there, but you could also enter this equation into the OSC (using x instead of c), and you'll see that x is equal to 50.4. (You might need to look around the graph, but there it is, a vertical line at 50.4.)

4. During an experiment, researchers determined that the equation $n = -\frac{3}{5}h + \frac{53}{5}$ can be used to model the number of bacteria in thousands (n) remaining in a Petri dish h hours after the start of the experiment. If the experiment was begun at 12:00 p.m., how many more bacteria remained in the dish the next day at 4:00 a.m. than remained in the dish earlier at 2:20 a.m.?

We have a linear equation with 2 unknowns but are then given times that allow us to plug in values for h. At 4:00 a.m., 16 hours have passed. Using 16 for h, we get $n = -\frac{3}{5}(16) + \frac{53}{5}$. This simplifies to $n = 1$. Let's not forget that n represents the number of bacteria *in thousands*, so that's **1,000**. At 2:20 a.m., 14 and 1/3 hours have passed, or $\frac{43}{3}$ hours (20 minutes is 1/3 of an hour). If we use $\frac{43}{3}$ for h, we get $n = -\frac{3}{5}(\frac{43}{3}) + \frac{53}{5}$. This isn't so bad once we cancel those 3's (or use the calculator), giving us $= -\frac{43}{5} + \frac{53}{5}$, and this simplifies to $n = 2$, which is **2,000**. That means the difference is **1,000**.

5. The average selling price of a house in a neighborhood of 30 houses is d dollars. The average selling price of a house in a nearby neighborhood of 25 houses is $12,500 less than that. Which of the following represents the average selling price of all the houses?

A) $55d - 12,500$
B) $\frac{d}{30} + \frac{d-12,500}{25}$
C) $\frac{11d - 62500}{11}$
D) $d - 62,500$

Drill Explanations | 187

As we so often do on this test, we'll use our average formula. If the average is d and the number of houses is 30, then the **sum** of the selling prices of those houses is $30d$. For the other houses, the sum is $25(d - 12,500)$. The grand sum of both sets of houses is $30d + 25(d - 12500)$ and we now need to divide that by the total number of houses, which is 55. Once we distribute the 25, we get $\frac{30d + 25d - 312,500}{55}$. This simplifies to $\frac{55d - 312,500}{55}$, and if we divide top and bottom by 5, we get **choice C**.

6. The equation $1 + st = s + t$ has infinite solutions for s if which of the following is true?

A) $s = -t$
B) $st = 2$
C) $t = 1$
D) $s = t$

If this equation has infinite solutions, then any numbers will fit for the unknown value. Once we try choice C, which is $t = 1$, we get $1 + s = s + 1$. Any value for s will work, so there are infinite solutions. The other choices do not allow for infinite solutions. Therefore, the answer is **choice C**.

7.
$$3y = \frac{3x}{2} - 9$$
$$2y = x$$

The equations above define 2 lines. How many intersecting points do they have?

A) 0
B) Exactly 1
C) Exactly 2
D) Infinitely many

One thing to notice is that choice C is impossible. Two lines cannot intersect 2 times. Can they have infinite intersection points? They can only if they are actually the same line. And we know that they have 0 intersecting points if they are parallel, which would mean they have the same slope.

This can be handled algebraically or with the OSC. Algebraically, you could get both equations into the form $y = mx + b$. The top equation gives you $y = \frac{1}{2}x - 3$, and the other equation is simply $y = \frac{1}{2}x$. They're different lines since they have different y-intercepts, but they do have the same slope of ½. They therefore do not intersect, and the answer is **choice A**.

With the OSC, you could simply enter the equations exactly as given. The only problem is that they will certainly look parallel, but it's impossible to say just by looking that they are precisely parallel (although it's a pretty safe bet).

Day 17

Advanced Geometry – Geo w/Trig, Arcs, Translations

After reading this material and completing the drill at the end, sign and check the box for Day 17.

The trickier geometry questions will likely still rely on the basics you've learned earlier in this course. To make the questions a little tougher, the test writers will include several of these basic ideas, throw in some trig, or include some not-that-simple algebra or arithmetic. But there are a few geometry concepts we haven't yet covered. We'll look at them all, but first, here's one with a little **trig** thrown in.

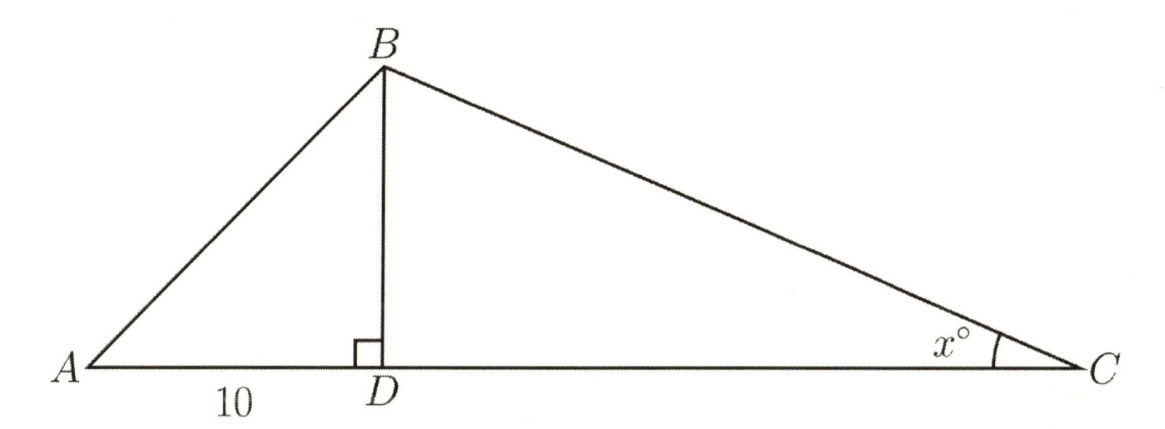

In the figure above, $\overline{AD}$ is 10 units long, and $\overline{BC} - \overline{DC} = 2$. If $\cos x = \frac{12}{13}$, how many coordinate units long is $\overline{AB}$?

A) 5
B) 10
C) $10\sqrt{2}$
D) $10\sqrt{3}$

Let's start with the trig. The ratio 12 : 13 should ring a bell. One of the special right triangles to remember is the 5 : 12 : 13. In this case, the actual lengths of the 2 sides can't be 12 and 13 since we're also told that their difference is 2. That means their actual lengths are 24 and 26. Furthermore, this lets us conclude that the other leg, BD, is 10. It's a 5 : 12 : 13 triangle with those numbers doubled.

Since AD is also 10, we know that we have another special right triangle—the 45/45 right isosceles triangle. With legs of 10, the hypotenuse is **10√2**. Know those specials!

But also—

If the figure is drawn to scale, then we can see that the side we want is a little more than 10. That certainly knocks off the first 2 answer choices. The test people say yes—**all figures are drawn to scale**—but just in case, be on the lookout for a figure that says, "Figure not drawn to scale." That might still show up.

We'll look at 1 now that tests a rule you *should* know as well as some algebra.

A regular 9-sided polygon is altered so 1 of the angles is twice the measure of 1 of the other angles. No other angles are altered. What is the measure of the smaller of those 2 angles?

You must know that the sum of all the angles in an *n*-sided polygon is $(n-2)180$ degrees. In this case, that works out to be 7 times 180, which is 1260 degrees. For a regular polygon, the angles are all the same, and if we divide 1260 by 9, we get 140 degrees for each angle. Since only 2 angles are altered, the sum of those 2 is still 140 x 2, which is 280 degrees. Now we can use our algebra: $x + 2x = 280$. That works out to be $x = 93\frac{1}{3}$ degrees. That's the smaller of the 2 angles, so that's the answer. If this was a student-produced response question without answer choices, **you can't simply enter 9313**. You could enter **93.33** or **280/3**.

More about Circles

You should, of course, know the relationship between the 4 elements of a circle: radius, diameter, circumference, and area. And you should know that the measure of a circle's angle is 360 degrees. That means a semicircle would be 180 degrees. We can continue with this idea.

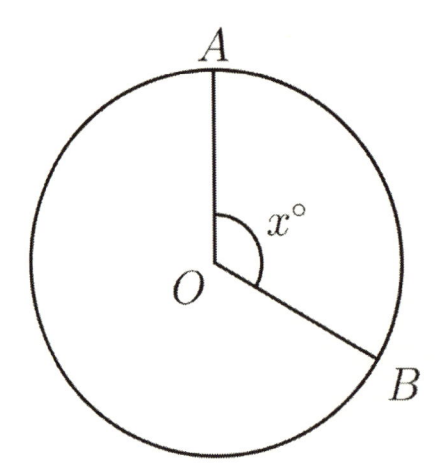

Angle *AOB* is referred to as a central angle of the circle. In this case, its measure is *x* degrees. Arc *AB* is "created" by, or *intercepted* by, this angle, and it has the same measure—it's also *x* degrees. To be clear, this arc can be measured in degrees, as we just said, but it also has a length. The relationship between these 2 is as follows. The length is a fraction of the circumference. What fraction? It's whatever fraction that central angle is **of 360**. If x is 90, then since that's ¼ of 360, the length of the arc would be ¼ of the circumference. Let's use this figure to create a question.

In the figure above, angle *AOB* is a central angle of circle *O*. If the radius of the circle is 9 and the length of arc *AB* is 5π, what is the value of *x*?

(Note: The notation "arc *AB*" refers to the short distance around the circumference between those 2 points. This is sometimes referred to as *minor arc AB*, and the long way around would be *major arc AB*. If it doesn't say minor or major, it's minor.)

Since we know the radius of the circle, we know everything about it. What we're interested in is the circumference. With a radius of 9, the circumference is 18π. What fraction of that is taken up by the 5π arc? That would be $\frac{5\pi}{18\pi}$, which simplifies to $\frac{5}{18}$. Therefore angle x must be $\frac{5}{18}$ of 360, which is **100**.

On a related note, have a look at this figure.

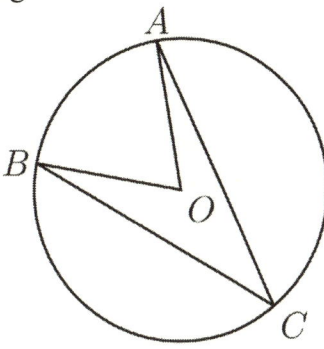

Similar to what we just looked at, the central angle *AOB* has the same measure as arc *AB*. If the central angle is 80 degrees, then arc *AB* is 80 degrees. But the measure of angle *ACB* is only half of that, or 40 degrees. So if you have an angle with 1 point on the circumference that intercepts an arc on the "other side" of the circle, its measure is half the arc angle.

Here's another little something about circles that could be tested. When a line touches a circle at exactly 1 point along its circumference, we say the line is *tangent* to the circumference. (This is not directly related to the trig "tangent.") And **the angle formed by this line and a radius of the circle is a right angle**. Here's an example and a question.

Advanced Geometry – Geo w/Trig, Arcs, Translations | 191

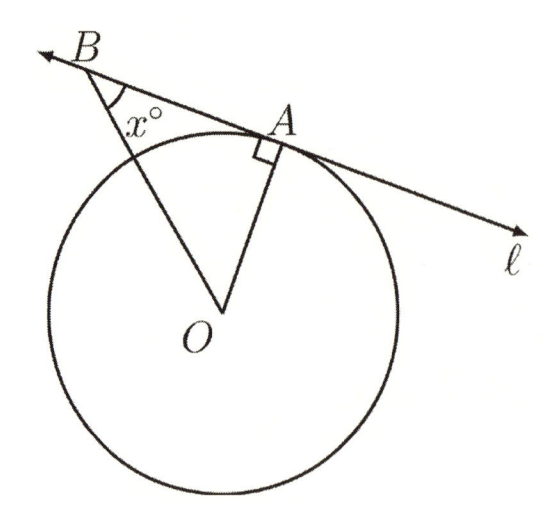

In the given figure, line *l* is tangent to the circle at point *A*, and point *B* lies on the line. If central angle *AOB* is 85 degrees, what is the value of *x*?

This shouldn't take long. Because of the rule stated above, we know that this is a right triangle, with the 90 degrees where the radius and line touch. Since the central angle is 85 degrees, *x* must equal **5.**

3 Dimensions

Let's put on our 3D glasses. You know how to find the volume of a rectangular solid (a box), right? It's just the product of the 3 dimensions—length, width, and height. And the surface area is just the sum of the 6 surfaces. But what about the diagonal? You can think of this as the longest line segment you could draw within the solid, like from the top, left, back corner to the bottom right, front corner (or vertex), as shown below with the dashed line.

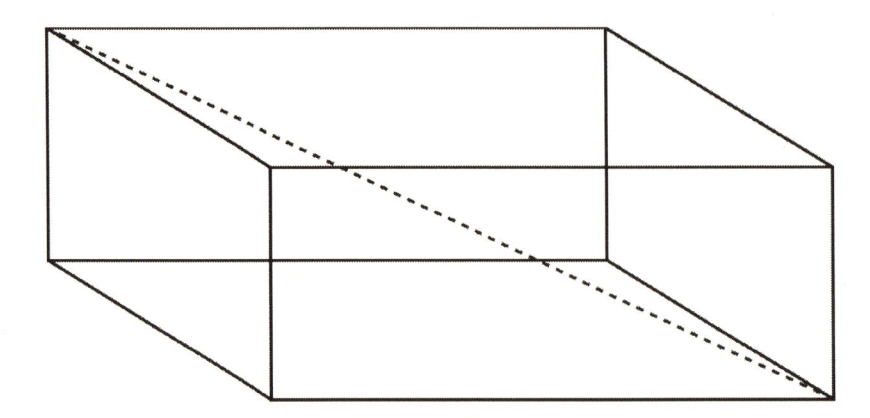

There's a simple way to calculate this length, which you can think of as the 3D version of the Pythagorean theorem: $d^2 = l^2 + w^2 + h^2$. In other words, **the diagonal squared is equal to the sum of the squares of each of the 3 dimensions**—length, width, and height.

Translations

As you probably know, geometric shapes can be moved around, or *translated*, within the coordinate plane. A **line**, for example, can be translated up, down, or side to side, which will affect the line's x and y intercepts, though not its slope.

A line in the xy-coordinate plane is defined by the equation $-3y + 4 = x$. If this line is translated down 2 units, what is the new y-intercept?

Now that we're in the coordinate plane, we're more likely to use the OSC. Since we're interested in the y-intercept, let's find out what it is *before* it's translated down. We can do that by setting x equal to 0 or by graphing the line and reading the y-intercept off the graph. If we do the algebra, we'll get $\frac{4}{3}$. If we look at the graph, we'll see 1.333, which you should recognize as the decimal approximation of this same value. When we translate down 2 units, the y-intercept drops 2, making it $\frac{4}{3} - 2 = -\frac{2}{3}$.

What if we translate this same line to the left 3 units? What would be the new x-intercept? Again, we can do this algebraically or with the OSC. Algebraically, we just let y equal 0, giving us an x-intercept of 4. Since this line gets translated to the left 3 units, the new x-intercept is **1**.

Using the OSC, we can easily see that the original x-intercept is 4 and then arrive at 1 the same way.

Note: There might be some confusion about moving shapes to the left and right. In this last example, the shift to the left did lead us to subtract 3 from the x-intercept. But let's look at the new equation. It was $-3y + 4 = x$. We can get this into a more familiar form by isolating x, giving us $y = -\frac{1}{3}x + \frac{4}{3}$. The translation of 3 to the **left** gives us this new equation, $y = -\frac{1}{3}(x + 3) + \frac{4}{3}$. The leftward shift results in **adding** 3 to the x value. Naturally, a rightward shift would do the opposite.

Parabolas can be translated up, down, left, and right, as well as upside down. Let's think about the "basic" parabola, the one with the equation $y = x^2$. This is an upward opening parabola that has a vertex at $(0, 0)$. Here's how shifts (translations) affect the equation:

Shifted up 2: $y = x^2 + 2$
Shifted down 2: $= x^2 - 2$
Shifted to the left 2: $y = (x + 2)^2$
Shifted to the right 2: $y = (x - 2)^2$
Turned upside down: $y = -x^2$

Also, in the equation $y = x^2$, the coefficient of x^2 is 1. If this is increased to 2, 3, and so on, the parabola narrows. Conversely, if the coefficient is decreased to ½, ¼, and so on, the parabola widens.

Translating the position of circles shouldn't be a problem as long as you've mastered the circle-defining equation—$(x - h)^2 + (y - k)^2 = r^2$—where (h, k) is the location of the center point, and r is the radius length. If we have the circle defined by $(x - 3)^2 + (y - 4)^2 = 100$ and want to shift it 7 units to the right, we end up with $(x - 10)^2 + (y - 4)^2 = 100$.

If we want to shift it up 5, we would end up with $(x-10)^2 + (y-9)^2 = 100$. For many of us, these shifts seem the reverse of what we would expect, so be careful!

Reflecting a shape involves moving it so it forms a reflected image of the original shape over a **mirror line**. This mirror line is often one of the axes. Here's a rectangle in Quadrant II of the *xy* plane.

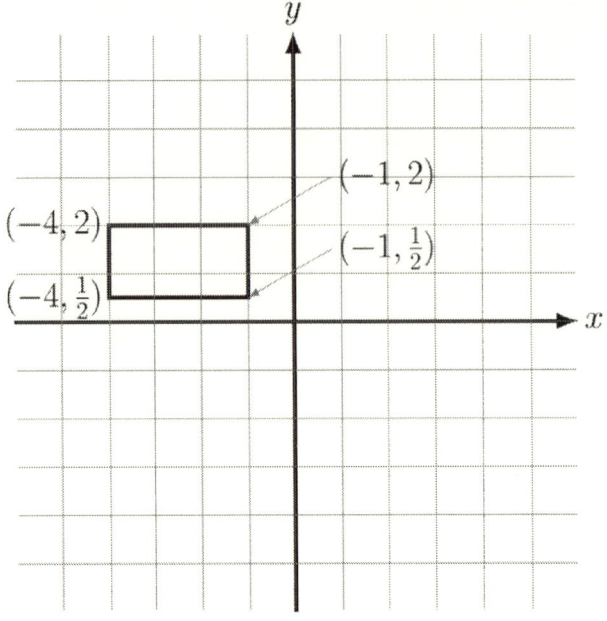

If we reflect this over the *x*-axis, we flip it *over* the axis, giving us this:

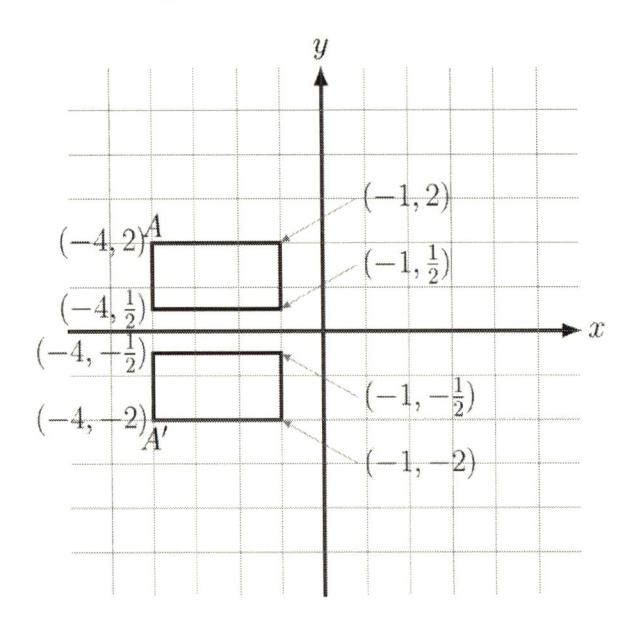

In this example, the vertex labeled *A* has been reflected to vertex *A'* called "*A* prime." You'll notice that the *x*-coordinates of each pair of reflected points stays the same (they're both -4) while their *y*-coordinates are opposites (one is 2 while the other is -2). If we had reflected our original rectangle over the *y*-axis, the opposite would happen.

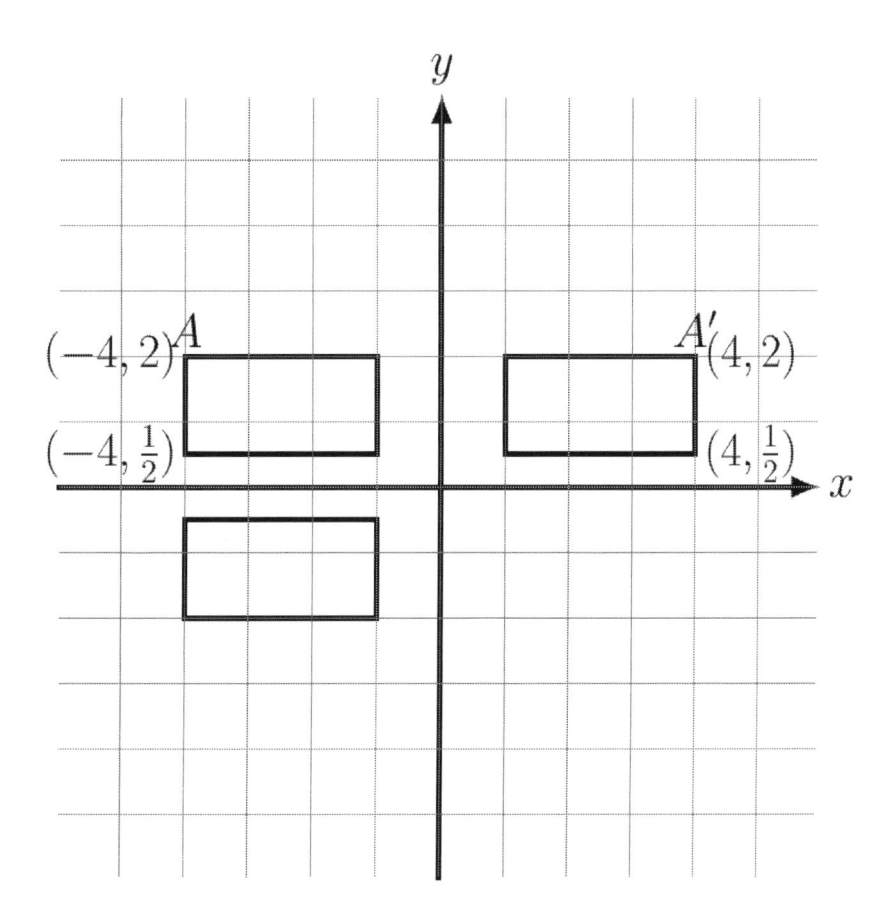

Other shapes can be reflected, of course, including points. Remember that when the reflection is over the *x*-axis, the *x*-coordinates stay the same, and the *y*-coordinates flip to the opposite sign. The reverse is true when the reflection is over the *y*-axis.

Day 17

Drill

Complete all the questions below.

1.

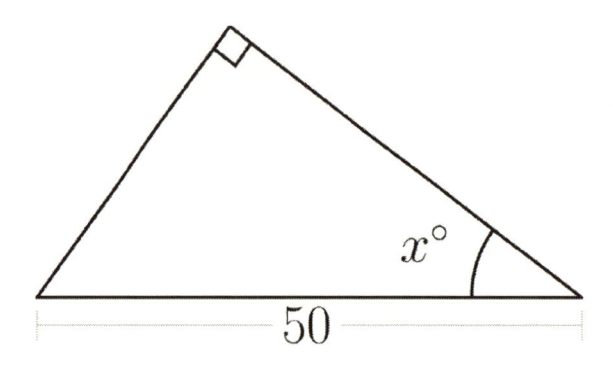

In the triangle shown above, $\cos x = 0.8$. What is the perimeter of the triangle?

2.

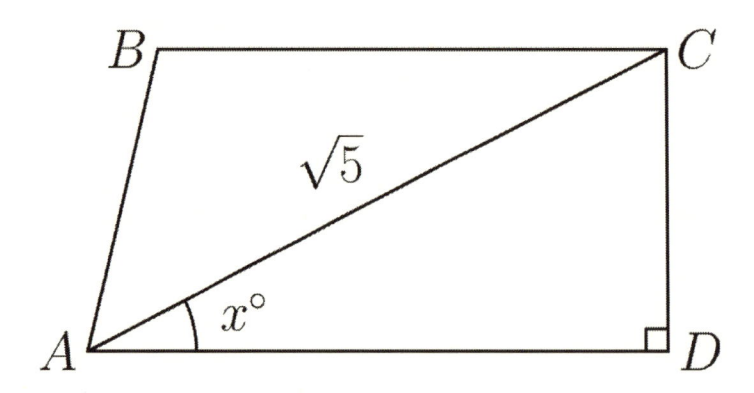

The figure shown above is right trapezoid $ABCD$. The length of $\overline{AC}$ is $\sqrt{5}$ inches. If $\tan x = \frac{1}{2}$ and the area of the trapezoid is $1\frac{7}{8}$ inches, what is the length of $\overline{BC}$?

3.

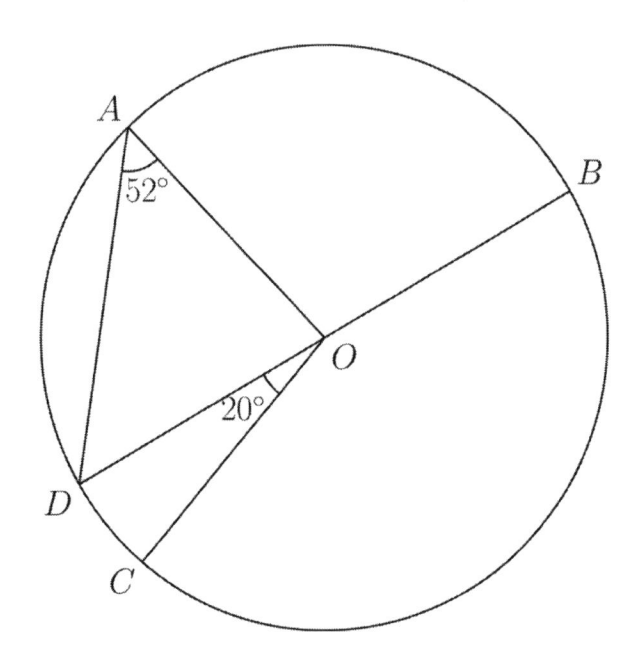

In the figure shown above, points A, B, C, and D lie on the circumference of circle O, which has a radius of 6. Which of the following is equal to the length of arc AB minus the length of arc DC?

A) $\dfrac{14\pi}{5}$
B) $\dfrac{10\pi}{9}$
C) $\dfrac{8\pi}{9}$
D) $\dfrac{5\pi}{9}$

4.

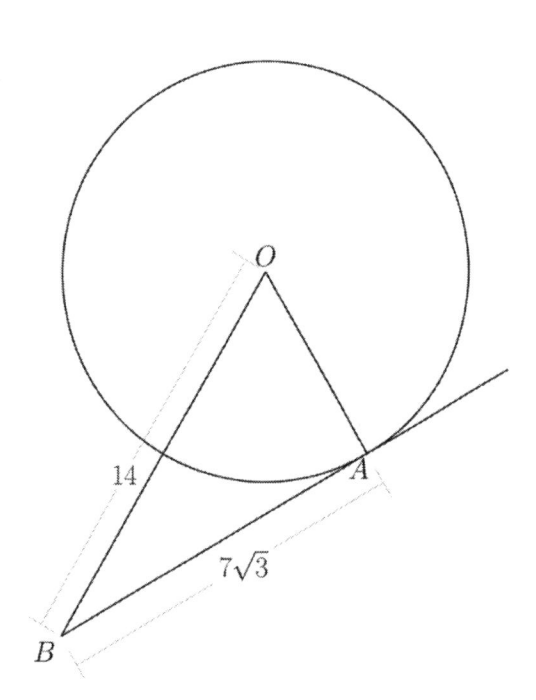

Drill | 197

In the figure above, points A and B lie on a line segment that is tangent to circle O at point A. What is the area of the circle in square units?

A) 7π
B) 14
C) 14π
D) 49π

5. A cube has a volume of 27 cubic inches. What is the length of the longest line segment that can be drawn within the cube in cubic inches?

A) 3
B) $3\sqrt{3}$
C) 6
D) $9\sqrt{3}$

6. Line m is defined by the equation $-y - 11 = -2x$ in the xy-coordinate plane. Which of the following equations defines the line after it has been translated down 6 units?

A) $-y - 11 = -8x$
B) $-6y - 11 = -2x$
C) $\frac{y}{2} = x - 8.5$
D) $\frac{y}{2} + x = 8.5$

7. In the xy-plane, the function f is defined by $f(x) = (x-1)^2 - 2$. This function has been translated to create the function g, which is defined by $g(x) = (x-4)^2 - 1$. Which of the following describes the translation from f to g?

A) 1 unit down and 3 units to the left
B) 1 unit up and 3 units to the left
C) 1 unit down and 3 units to the right
D) 1 unit down and 3 units to the right

8. An upward opening parabola defined by the equation $f(x) = ax^2 + bx + c$ has 2 x-intercepts, where a, b, and c are non-0 constants. Which of the following is true about the x-intercepts of the parabola defined by the equation $g(x) = (a+1)x^2 + bx + c$?

A) They are the same as the x-intercepts of function f.
B) They are closer together than the x-intercepts of function f.
C) They are farther apart than the x-intercepts of function f.
D) They are each greater than the 2 x-intercepts of function f.

9. A circle is defined by the equation $(x+4)^2 + y^2 = 2$ in the xy-plane. If the circle is translated 1 unit up, 2 units to the left, and its radius is increased by 1 unit, what is the equation that would then define the circle?

A) $(x-6)^2 + (y-1)^2 = 4$
B) $(x-6)^2 + (y+1)^2 = 4$
C) $(x+6)^2 + (y+1)^2 = 4$
D) $(x+6)^2 + (y-1)^2 = 4$

10. Point Q is located at $(-10, 8)$ on the xy plane. If it is reflected over the y-axis and then over the x-axis, which of the following is the length in coordinate units between its original location and its location after the 2 reflections?

A) $2\sqrt{41}$
B) $4\sqrt{41}$
C) $32\sqrt{5}$
D) 656

Day 17

Drill Explanations

1.

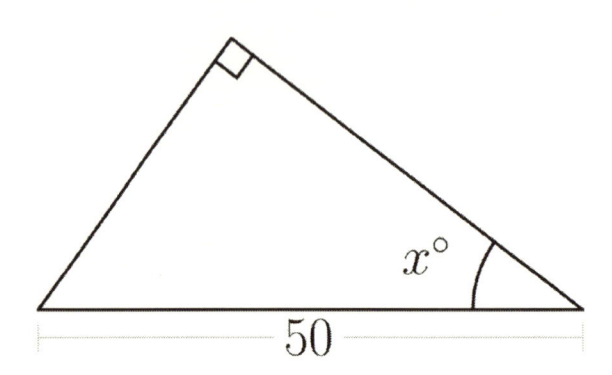

In the triangle shown above, cos x = 0.8. What is the perimeter of the triangle?

It's usually helpful to think of our trig ratios as fractions rather than decimals. Since 0.8 is equal to 4/5, we know that the ratio of the side that's adjacent to x **to** 50 is 4 : 5. That means the side adjacent to x is 40. Once you see those numbers in a right triangle, you should think "3 : 4 : 5 special." Without having to use the Pythagorean theorem, we know that the side opposite x is 30, and the perimeter is 30 + 40 + 50 = 120.

2.

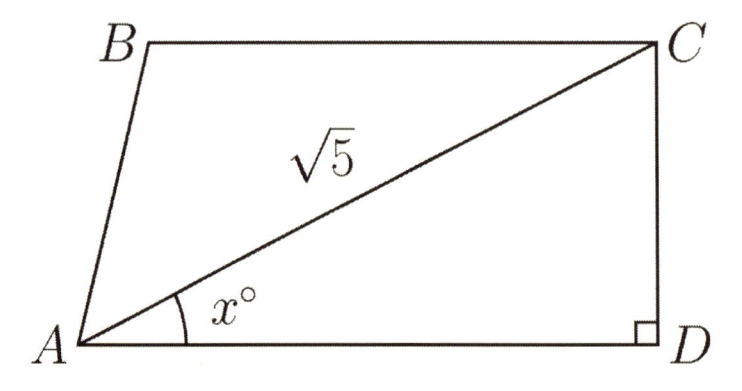

The figure shown above is right trapezoid $ABCD$. The length of $\overline{AC}$ is $\sqrt{5}$ inches. If $\tan x = \frac{1}{2}$ and the area of the trapezoid is $1\frac{7}{8}$ square inches, what is the length of $\overline{BC}$?

From the tangent information, we know that the ratio of the side opposite x (side CD) to the side adjacent to x (side AD) is 1 : 2. This by itself doesn't mean that the actual lengths are 1 inch and 2

200 |Day 17

inches, but in this case, they are. You might realize that using the Pythagorean theorem, we get $1^2 + 2^2 = \sqrt{5}^2$, or simply $1 + 4 = 5$. If that didn't occur to you, you can set up the equation $a^2 + (2a)^2 = \sqrt{5}^2$ since we know that 1 leg is twice the length of the other. That also gives us the values 1 and 2 for the leg lengths.

We're almost there. Let's call the length of $\overline{BC}$ x (that's what we're looking for). We're given the area of the trapezoid as $1\frac{7}{8}$ square inches, so knowing the formula for the area of a trapezoid, we can say $\frac{x+2}{2} \times 1 = 1\frac{7}{8}$. Let's simplify this to $\frac{x+2}{2} = \frac{15}{8}$. With some cross-multiplication, we get $8x + 16 = 30$ and eventually $x = \frac{7}{4}$, or 1.75 inches.

3.

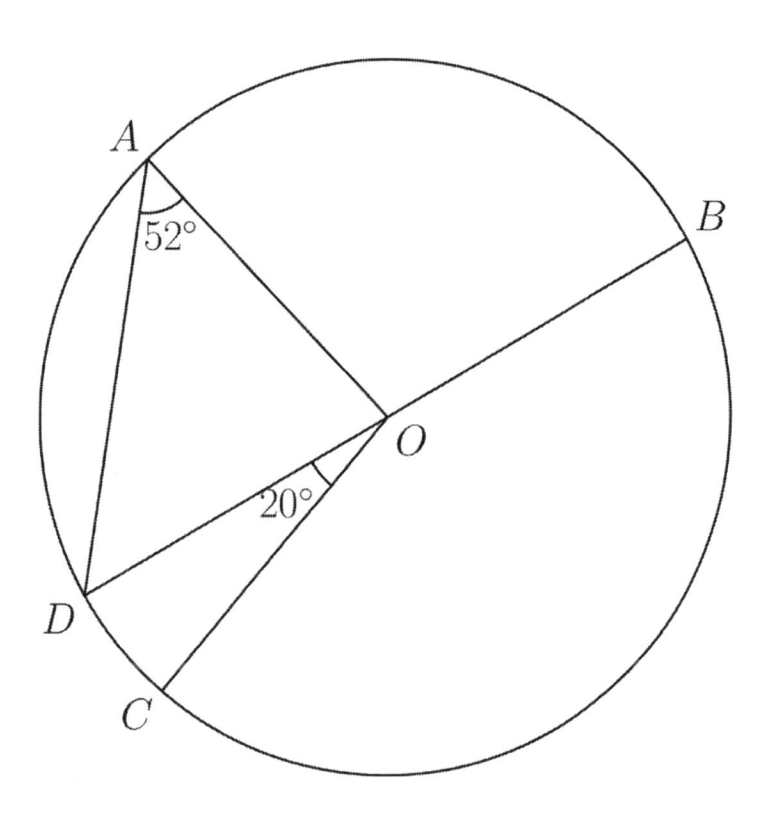

In the figure shown above, points *A, B, C,* and *D* lie on the circumference of circle *O*, which has a radius of 6. Which of the following is equal to the length of arc *AB* minus the length of arc *DC*?

A) $\frac{14\pi}{5}$
B) $\frac{10\pi}{9}$
C) $\frac{8\pi}{9}$
D) $\frac{5\pi}{9}$

Since we're interested in arc lengths, let's first determine the circumference. With a radius of 6, that would be 12π.

If we now look at triangle *AOD*, we see that 2 of the sides are radii, so this is an isosceles triangle. That means angle *ADO* must also be 52 degrees. Since arc *AB* is intercepted by (created by) this angle *and* since point *D* is on the circumference, arc *AB* is twice 52 degrees, or 104 degrees. (That's its measure in *degrees* but not its *length*.)

We can get its length now, but let's first get the angle measurement of arc *DC*. This is intercepted by a **central** angle of 20 degrees, so the arc is also 20 degrees. If we subtract this—since that's what the question is about—from the other arc, we get $104 - 20 = 84$. So the difference, **in degrees**, is 84. What length is that? Since the entire circle is 360 degrees, we need to find $\frac{84}{360}$ of 12π. Since the answer choices have π, we don't have to approximate the answer; π will just stay π. You could do the calculation yourself using cancellation to reduce the numbers, or you could use the OSC. If you just do $\frac{84}{360}$ times 12 on the OSC, you'll get 2.8, which is the same as $\frac{14}{5}$, giving us **choice A**. (If you don't see that as obvious, you could use the decimal-to-fraction function on the OSC on the left of where you enter numbers.)

4.

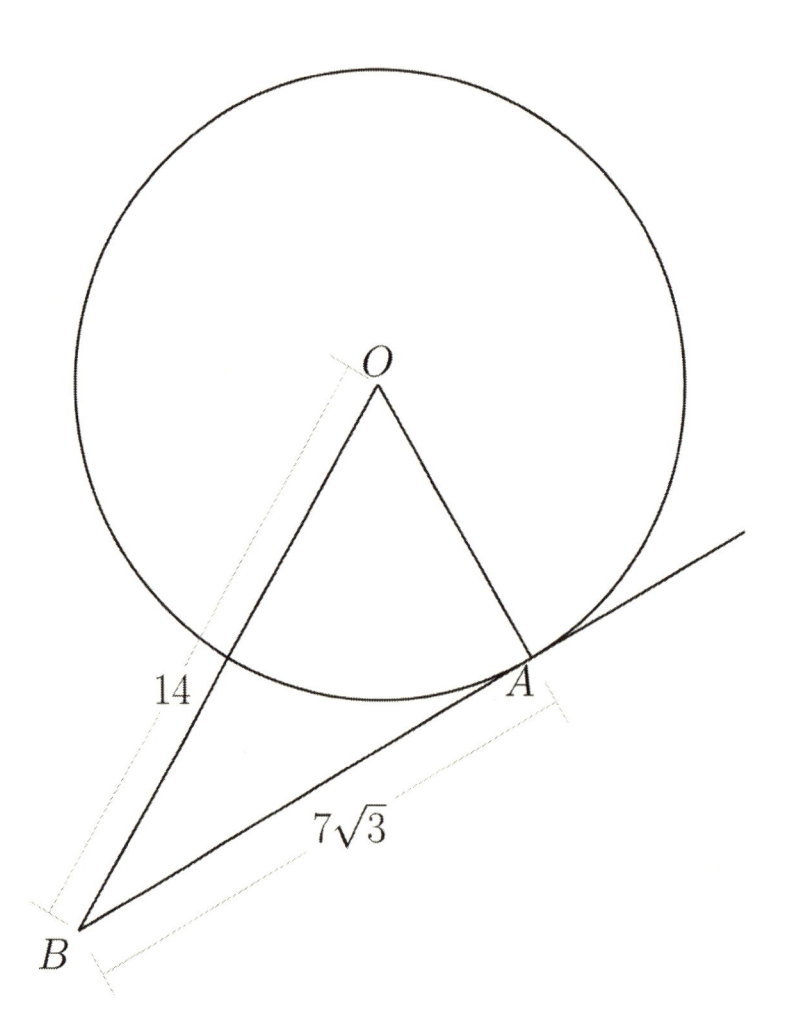

In the figure above, points *A* and *B* lie on a line segment that is tangent to circle *O* at point *A*. What is the area of the circle in square units?

A) 7π
B) 14
C) 14π
D) 49π

Whenever you see a line tangent to a circle, you should realize that they form a right angle at their intersection. We now know that AOB is a right triangle. Since we're always on the lookout for the special right triangles, you should also note that this is a 30/60. The clue is that the longer leg has $\sqrt{3}$ in it. That alone doesn't tell us that it's a 30/60, but since the hypotenuse is 14, you should recognize this as a $7, 7\sqrt{3}, 14$ triangle—in other words, a 30/60. Furthermore, OA is not only a leg of the triangle; it's a radius. So the area is 49π, which is **choice D**.

5. A cube has a volume of 27 cubic inches. What is the length of the longest line segment that can be drawn within the cube in cubic inches?

A) 3
B) $3\sqrt{3}$
C) 6
D) $9\sqrt{3}$

This question is simply asking for the diagonal. With a volume of 27 cubic inches, any edge of this cube is 3. Using the diagonal formula, we then have $d^2 = 3^2 + 3^2 + 3^2$. That gives us $d^2 = 27$, so the diagonal is $\sqrt{27}$, which simplifies to $3\sqrt{3}$, which is **choice B**.

6. Line *m* is defined by the equation $-y - 11 = -2x$ in the *xy*-coordinate plane. Which of the following equations defines the line after it has been translated down 6 units?

A) $-y - 11 = -8x$
B) $-6y - 11 = -2x$
C) $\frac{y}{2} = x - 8.5$
D) $\frac{y}{2} + x = 8.5$

This can be solved algebraically without much trouble or with a little help from the OSC. Algebraically, we can get this equation into the typical form with *y* on one side—$y = 2x - 11$. If a line is translated down 6, then the *y*-intercept decreases by 6, giving us $y = 2x - 17$. None of the choices look like this, but if you double both sides of **choice C** in order to isolate *y*, that's what you'll get. (You didn't really need to re-form the original equation as long as you're sure that the intercept is -11 and not, for example, positive 11.)

With the OSC, you can enter the equation as is and then enter the others (perhaps using the duplicate button for efficiency) until you see which one looks like the original has been translated down 6 units. Neither method should take long.

Drill Explanations | 203

7. In the *xy*-plane, the function *f* is defined by $f(x) = (x-1)^2 - 2$. This function has been translated to create the function *g*, which is defined by $g(x) = (x-4)^2 - 1$. Which of the following describes the translation from *f* to *g*?

A) 1 unit down and 3 units to the left
B) 1 unit up and 3 units to the left
C) 1 unit down and 3 units to the right
D) 1 unit up and 3 units to the right

The change from $(x-1)$ to $(x-4)$ is a rightward shift of 3. The change from -2 to -1 is an upward shift of 1. That leads us to **choice D**.

8. An upward opening parabola defined by the equation $f(x) = ax^2 + bx + c$ where *a*, *b*, and *c* are positive constants has 2 *x*-intercepts. Which of the following is true about the *x*-intercepts of the parabola defined by the equation $g(x) = (a+1)x^2 + bx + c$?

A) They are the same as the *x*-intercepts of function *f*.
B) They are closer together than the *x*-intercepts of function *f*.
C) They are farther apart than the *x*-intercepts of function *f*.
D) They are each greater than the 2 *x*-intercepts of function *f*.

By increasing the coefficient of the x^2 term, the parabola narrows, moving the intercepts closer together, as it says in **choice B**.

9. A circle is defined by the equation $(x+4)^2 + y^2 = 4$ in the *xy*-plane. If the circle is translated 1 unit up and 2 units to the left, and if its radius is increased by 1 unit, what is the equation that would then define the circle?

A) $(x-6)^2 + (y-1)^2 = 5$
B) $(x-6)^2 + (y+1)^2 = 5$
C) $(x-6)^2 + (y-1)^2 = 9$
D) $(x+6)^2 + (y-1)^2 = 9$

You can efficiently eliminate 2 choices by realizing that since the constant is 4, the original circle has a radius of 2. If we increase that to 3, the constant term should be 9. That eliminates the first 2 choices. Of the remaining choices, both correctly subtract 1 from *y* in order to translate the circle up 1, but translating the circle to the left by 2 **adds** 2 to the original value of 4, giving us **choice D**.

10. Point *Q* is located at $(-10, 8)$ on the *xy* plane. If it is reflected over the *y*-axis and then over the *x*-axis, which of the following is the length in coordinate units between its original location and its location after the 2 reflections?

A) $2\sqrt{41}$
B) $4\sqrt{41}$
C) $32\sqrt{5}$
D) 656

The reflection of the point over the *y*-axis keeps the same *y*-coordinate but flips the sign of the *x*, so we get (10, 8). Flipping this over the *x*-axis changes only the sign of the *y*-coordinate, giving us (10, −8).

Now we need the distance between (−10, 8) and (10, −8). If you know the formula for the distance between 2 points, that's great. However, it's often efficient to use the idea behind the formula, which is simply the Pythagorean theorem. If we connect these 2 points—whether you sketch it out or just think about it—and then triangulate by "traveling" down 20 and across 16, we get a right triangle.

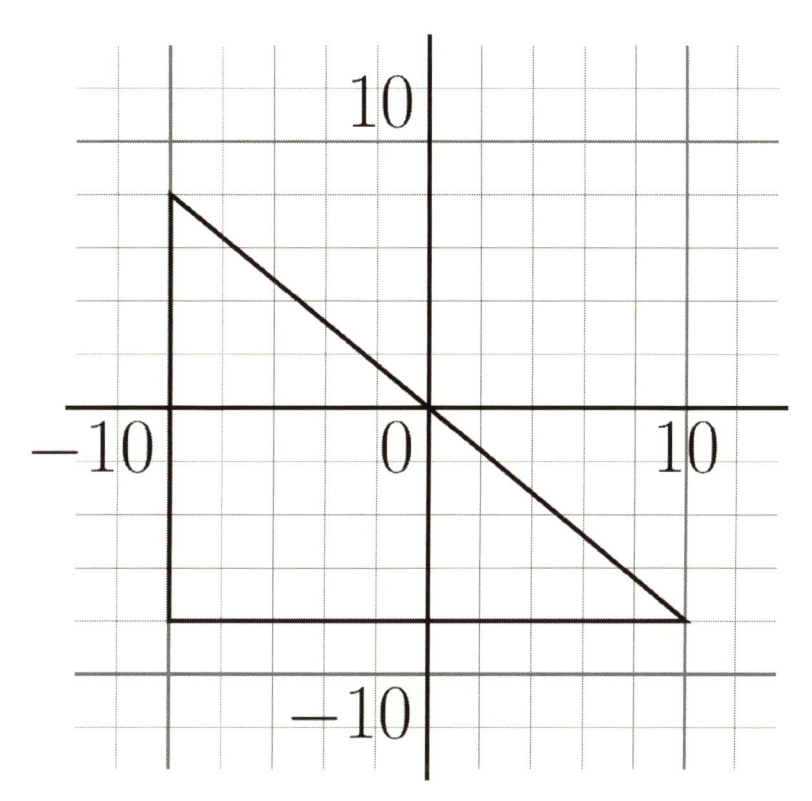

Is this a special (you should ask yourself)? No, not this time. So you can use the OSC to calculate *c* in the equation $20^2 + 16^2 = c^2$. That gives us $656 = c^2$, so the answer is $\sqrt{656}$, which, since it's an irrational number, the OSC gives us as a decimal. This is not obviously 1 of the choices (though we can now eliminate choice D). You could factor $\sqrt{656}$ so it matches 1 of the choices, or you could enter these choices into the calculator until you get that same decimal approximation. The answer is **choice B**.

Day 18

Absolute Value, Matching Coefficients, Sequences, Work Rate, Probability II, Statistical Inferences

After reading this material and completing the drill at the end, sign and check the box for Day 18.

We've now covered all the major and minor topics that will be on the test. Today we'll look at the kinds of questions we see on the test now and including—possibly—on *your* test. Let's be prepared for **anything**.

Absolute Value

Absolute value is a fairly simple concept, but the test folks are good at coming up with tricky ways to test it. Officially, the absolute value of a number is how far it is from 0 on the number line. Since 3 and negative 3 are just as far from 0, they have the same absolute value. In other words, $|3| = 3$, and $|-3| = 3$. We say this as "absolute 3 equals 3, and absolute negative 3 equals 3.

Let's throw in some algebra. If $|x| = 10$, then there are 2 solutions: $x = 10$ and $x = -10$, or simply $x = \pm 10$. The absolute value of an unknown will *always* have 2 solutions unless it turns out that the unknown is 0. Absolute 0 is simply 0.

Let's make it a bit more complicated. What if $|x - 4| = 10$? Like before, $x - 4 = 10$, and $x - 4 = -10$. Again, this gives us 2 solutions: $x = 16$ and $x = -4$. Make sure you totally get this.

Sometimes the test expresses it a little differently. Given $|x - 4| = 10$, then the 2 equations we can form are $x - 4 = 10$ and $-(x - 4) = 10$. Do you see how this is the same thing, still giving us the same 2 solutions?

Now let's combine absolute value with inequalities. Let's say you have $|r - 7| > 5$. With inequalities, we get a *range* of values, and now we'll get **2** of them. We can conclude that $r - 7$ can be any value greater than 5, but it also can be any value **less than** -5. For example, $r - 7$ (*not r, itself*) could be 6. These 2 ranges do not overlap. Let's solve them separately. If $r - 7 > 5$, then $r > 12$. If $r - 7 < -5$, then $r < 2$. Therefore, given $|r - 7| > 5$, we can say that $r > 12$ or $r < 2$.

Now what if $|r - 7| < 5$? One range of solutions is $r - 7 < 5$. That means $r < 12$. But we also have $r - 7 > -5$, and that means $r > 2$. These ranges *do* overlap, so we know that r is any number less than 12 **that is also** greater than 2. We express this as $2 < r < 12$.

Matching Coefficients

Watch out for the phrase "for all values of x." These questions usually involve, for example, having ax^3 on 1 side of an equation and $7x^3$ on the other side. We can say that those 2 coefficients are equal, which means $a = 7$ because they are the coefficients of **matching powers**. Let's look at a question.

$$6x^2 + sx + t = (3x + 1)(rx - 7)$$

In the given equation, r, s, and t are constants. If the equation is true for all values of x, what is the value of $r + s + t$?

First, we'll distribute the terms on the right. The equation is now $6x^2 + sx + t = 3rx^2 + rx - 21x - 7$. As in the description above, the coefficients of $6x^2$ and $3rx^2$ are equal. That means $r = 2$. What's next? **Remember to factor!** We can factor an x out of the middle 2 terms on the right. While we're at it, we'll subtract $6x^2$ from each side. They're gone. Knowing that r is equal to 2, we now have $sx + t = x(2 - 21) - 7$, which simplifies to $sx + t = -19x - 7$. Since we again have matching powers on each side, we know that $s = -19$, and we know that $t = -7$. Adding our 3 constants, we get $2 + (-19) + (-7) = -24$.

Sequences

There are 2 basic kinds of sequences you might see on the test—arithmetic and geometric. In an arithmetic sequence, consecutive numbers are separated by a common difference. Examples are $(-3, -2, -1, 0, 1, 2)$, $(5, 11, 17, 23, 29)$, and $(100, 90, 80, 70, 60)$. In other words, you add/subtract the same number each time. Here's a sample question.

In an arithmetic sequence, $n_1 = -6$ and $n_4 = 9$. What is the value of n_6?

The notation n_1 refers to the 1st number in the sequence. n_4 is the 4th number in the sequence. One way to approach this is to see that we added 15 to get from the 1st to the 4th term. That's an increase of 15 in 3 "jumps." That means each jump was a jump of 5. In other words, the common difference is 5. To get from the 4th term to the 6th we need 2 jumps of 5, so that jump of 10 brings us to 19.

Let's say you want the 36th term in this sequence. Starting with the 1st term, which is -6, we would have to add 5 to this number 35 times. That's like saying we would have to add 5×35, which is 175. Adding that to -6, we get 169.

In a geometric sequence, consecutive numbers are separated by a common ratio. Examples are $(2, 4, 8, 16, 32)$, $(-10, -30, -90, -270)$, and $(1, -5, 25, -125, 625)$. In other words, you multiply/divide by the same number each time. And so—

In a geometric sequence, $n_5 = 48$ and $n_6 = 192$. What is the value of n_1?

To advance from the 5th to the 6th term, we multiply by 4. To go back to the 1st term, we need to divide the 5th term (48) by 4, **4 times**. In other words, we need to divide 48 by 4^4. Using the OSC, if we divide 48 by 4^4, we'll get 0.1875. If the answer needs to be in fractional form, you can hit the fraction button on the calculator to get $\frac{3}{16}$.

Work Rate

We briefly looked at a question of this sort before. If you have 2 people (or machines or whatever) working at the same time at different rates, you can find their combined time by dividing the product of their times by the sum of their times. Let's look at an example.

Amir and Basil work together in a furniture store that sells tables that need to be assembled. Amir can assemble a table in 6 minutes. Basil can assemble a table in 7 minutes. Working together and simultaneously at these rates, how many minutes does it take them to completely assemble 20 tables? (Round your answer to the nearest integer.)

Again, we want $\frac{product}{rate}$ to get combined time. In this case, we'll get $\frac{6 \times 7}{6+7} = \frac{42}{13}$. So their combined time is $\frac{42}{13}$ minutes for 1 table. The question asks how long it takes them to completely assemble 20 tables, so we'll multiply this time by 20, giving us $\frac{840}{13}$ minutes, which is a little greater than 64.6 minutes. Therefore, the answer is **65**.

Let's now include a 3rd coworker, Anusha, who takes 12 minutes to assemble a table. Now all 3 workers together will assemble tables. We can't simply divide the product of all 3 times by their sum to get the combined time. We **can** do that for 2 of them as we've done already and then calculate product/sum using this combined time and the 3rd person's time (12 minutes).

However, there is another approach that works for any number of times. The formula looks like this:

$$\frac{1}{T_1} + \frac{1}{T_2} + \frac{1}{T_3} \ldots = \frac{1}{T_t}$$

What this means is that the sum of the *reciprocals* of the individual times is equal to the reciprocal of the total (combined) time. So—

$$\frac{1}{6} + \frac{1}{7} + \frac{1}{12} \ldots = \frac{1}{T_t}$$

This becomes $\frac{14+12+7}{84} = \frac{33}{84} = \frac{11}{28}$. It is the reciprocal of their combined time, which would then be $\frac{28}{11}$ minutes. That's for 1 table. For 20 tables, we multiply by 20 to get $\frac{560}{11}$, which is close to **51 minutes**.

Probability

We've looked at basic probability where we create a fraction with the total number of possibilities on the bottom and the "desired" number on top. That can then be converted to a percentage or decimal if necessary.

We might need to calculate the probability of more than 1 event. What matters is whether the outcome of 1 event affects the likelihood of the other event. For example, if you flip a coin, there's a ½ probability that it lands heads up. If you flip it again, there's still a ½ probability that it lands heads up. It doesn't matter what happened the 1st time. If you need to know the probability of flipping it 2 times and having it land heads up both times, you **multiply** the individual probabilities. Since $\frac{1}{2} \times \frac{1}{2} = \frac{1}{4}$, there's a $\frac{1}{4}$ probability of the coin landing heads up both times.

Let's say instead you have a deck of 52 playing cards, 4 of which are kings. The probability of randomly selecting a king from the deck is $\frac{4}{52}$. What is the probability of randomly selecting **another** king from the deck? There are only 3 kings left of the remaining 51 cards, so there is a $\frac{3}{51}$ probability of selecting that second king. To get the probability of selecting both kings, multiply $\frac{4}{52}$ by $\frac{3}{51}$ (which is $\frac{1}{221}$).

Statistical Inferences

What does statistical inferences mean? It means we must be careful making conclusions (inferences) from a sample (usually of people). Here's an example.

A radio station asked its listeners to tell them what song they think should be named Song of the Year in an upcoming national award show. Of the 1,779 people who contacted the station, more than 80% chose the song "Fools." Last year, the song chosen by most of the 1,698 people who contacted the station chose the song that won Song of the Year for that year. Based on these results, which of the following is the most appropriate conclusion?

A) The Song of the Year is usually the song that most people who listen to the radio station believe should win.
B) The people who listen to the radio station are representative of the people who decide which song will be named Song of the Year.
C) It's plausible that the song "Fools" will be named Song of the Year.
D) The number of people who will contact the station next year to say which song they believe should win Song of the Year will be greater than 1,779.

Absolute Value, Matching Coefficients, Sequences, Work Rate, Probability II, Statistical Inferences

This might not seem like math. It's logic, which is a close cousin of mathematics. In both cases, we can make conclusions **but** only if we have enough evidence. In this case, all the choices but **choice C** are drawing definite conclusions without enough information.

Day 18

Drill

Complete all the questions below.

1.
$$|x - 2| = -2x - 9$$

In the given equation, if $x < -3$, what is the value of x?

2. Which of the following inequalities is equivalent to $|r + 4| < 1$?

A) $-5 < r < -3$
B) $-3 < r < 5$
C) $-5 < r < 5$
D) $3 < r < 5$

3.
$$5ax^4 + bx^2 - cx + k = 15x^4 + (2x + 6)(x - 1) + 6$$

In the given equation, a, b, c, and k are constants. If the equation is true for all values of x, which of the following is true?

A) $b < a < c < k$
B) $k < b < c < a$
C) $c < k < b < a$
D) $k < b < a < c$

4. There are 10 winning tickets in a raffle out of a total of n tickets that are being sold. If Eva randomly buys 3 tickets, which of the following represents the probability that all of them are winning tickets?

A) $\dfrac{27}{n^3 - n^2 + 2n}$
B) $\dfrac{720}{n^3 - 3n^2 - 2n}$
C) $\dfrac{720}{n^3 - 3n^2 + 2n}$
D) $\dfrac{1000}{n^3 - 3n^2 + 2n}$

5. A group of 225 Millsburg residents over the age of 65 took part in a survey where they were asked whether they felt the town provided enough recreation opportunities for people their age. The majority of the people in this group answered "yes." At a recent public event, the mayor of Millsburg claimed that the results of the survey meant the town provided plenty of recreation opportunities for people over the age of 65. Which of the following pieces of information would most strengthen the mayor's claim?

A) The residents who took part in the survey were representative of Millsburg's residents over the age of 65.
B) The residents who took part in the survey did not know that the mayor was going to refer to the results of the survey.
C) The residents who took part in the survey had lived in Millsburg for at least 5 years.
D) Some of the residents who took part in the survey answered "not sure."

6. In an arithmetic sequence, $n_1 = 2$ and $n_3 = -10$. What is the value of n_{14}?

7. Computer A can perform a certain kind of complicated calculation in $\frac{1}{2}$ second. Computers B and C can do the same kind of calculation in $\frac{1}{3}$ and $\frac{2}{3}$ seconds, respectively. Working simultaneously, how long does it take the 3 computers to do 6 of these calculations?

Day 18

Drill Explanations

1.
$$|x - 2| = -2x - 9$$

In the given equation, if $x < -3$, what is the value of x?

The 2 equations we get are $x - 2 = -2x - 9$ and $x - 2 = -(-2x - 9)$. Solving the 1st one, we get $x = -\frac{7}{3}$. The other equation gives us $x = -11$. We're told that x is less than -3, so the answer is -11.

2. Which of the following inequalities is equivalent to $|r + 4| < 1$?

A) $-5 < r < -3$
B) $-3 < r < 5$
C) $-5 < r < 5$
D) $3 < r < 5$

The 2 inequalities we get are $r + 4 < 1$ and $r + 4 > -1$. These simplify to $r < -3$ and $r > -5$. When we put these together, we get **choice A**.

3.
$$5ax^4 + bx^2 - cx + k = 15x^4 + (2x + 6)(x - 1) + 6$$

In the given equation, a, b, c, and k are constants. If the equation is true for all values of x, which of the following is true?

A) $b < a < c < k$
B) $k < b < c < a$
C) $c < k < b < a$
D) $k < b < a < c$

When we expand (distribute) the right side of the equation and combine like terms, we get $15x^4 + (2x^2 + 4x - 6) + 6$, which further simplifies to $15x^4 + 2x^2 + 4x$. Therefore, we can rewrite the equation as follows:

$$5ax^4 + bx^2 - cx + k = 15x^4 + 2x^2 + 4x$$

Drill Explanations | 213

Since this equation is true "for all values of x," we know that the coefficients of matching powers are equal. This tells us that $a = 3$, $b = 2$, and $c = -4$, and since there is no constant by itself on the right side, $k = 0$. These values lead us to **choice C**.

4. There are 10 winning tickets in a raffle out of a total of n tickets that are being sold. If Eva randomly buys 3 tickets, which of the following represents the probability that all of them are winning tickets?

A) $\dfrac{27}{n^3 - n^2 + 2n}$
B) $\dfrac{720}{n^3 - 3n^2 - 2n}$
C) $\dfrac{720}{n^3 - 3n^2 + 2n}$
D) $\dfrac{1000}{n^3 - 3n^2 + 2n}$

The probability of selecting 1 winning ticket is $\dfrac{10}{n}$. After that, the probability of selecting another winning ticket is $\dfrac{9}{n-1}$. The probability of selecting a 3rd winning ticket is $\dfrac{8}{n-2}$. To get the probability of selecting all 3, we multiply these individual probabilities: $\dfrac{10}{n} \times \dfrac{9}{n-1} \times \dfrac{8}{n-2}$. Multiplying across the top and bottom, we get **choice C**.

5. A group of 225 Millsburg residents over the age of 65 took part in a survey where they were asked whether they felt the town provided enough recreation opportunities for people their age. The majority of the people in this group answered "yes." At a recent public event, the mayor of Millsburg claimed that the results of the survey meant that the town provided plenty of recreation opportunities for people over the age of 65. Which of the following pieces of information would most strengthen the mayor's claim?

A) The residents who took part in the survey were representative of Millsburg residents over the age of 65.
B) The residents who took part in the survey did not know that the mayor was going to refer to the results of the survey.
C) The residents who took part in the survey had lived in Millsburg for at least 5 years.
D) Some of the residents who took part in the survey answered "not sure."

A survey used to make a claim about a larger group can only be accurate if the group being surveyed is representative of the larger group. This is crucial for any survey, and this is exactly what **choice A** says.

6. In an arithmetic sequence, $n_1 = 2$, and $n_3 = -10$. What is the value of n_{14}?

Since the **1st term** in the sequence is 2 and the **3rd term** is -10, the common difference between consecutive terms in the sequence is -6. In other words, we subtract 6 to get to the next term. From the 1st term to the 14th term, we subtract 6 a total of 13 times from the 1st term. Since 13 times 6 is 78, we subtract 78 from 2 to get -76.

7. Computer A can perform a certain kind of complicated calculation in $\frac{1}{2}$ second. Computers B and C can do the same kind of calculation in $\frac{1}{3}$ and $\frac{2}{3}$ seconds, respectively. Working separately but simultaneously, how long does it take the 3 computers to do 6 of these calculations?

Since we have 3 individual times, we can determine the combined time using this formula:

$$\frac{1}{T_1} + \frac{1}{T_2} + \frac{1}{T_3} \ldots = \frac{1}{T_t}$$

Plugging in these values, we get the following:

$$\frac{2}{1} + \frac{3}{1} + \frac{3}{2} \ldots = \frac{1}{T_t}$$

This simplifies to $\frac{13}{2} = \frac{1}{T_t}$ and finally, T_t, which is the combined time, equal to $\frac{2}{13}$ seconds. That's how long it takes them to do 1 calculation. To do 6 of them, it would take 6 times this number, or $\frac{12}{13}$ seconds.

Day 19

Quiz 3

After reviewing days 12–18 and then completing Quiz 3, sign and check the box for Day 19.

Quiz 3

1.

$$\frac{\frac{5}{x^2-3}+\frac{y}{2x}}{\frac{1}{2}}$$

Which of the following is equivalent to the fraction given above?

A) $\frac{5x+x^2y-3y}{x^2-3x}$

B) $\frac{5x+2x^2y-6y}{x^3-3x}$

C) $\frac{10x+x^2y-3y}{x^3-x}$

D) $\frac{10x+x^2y-3y}{x^3-3x}$

2. In the equation $sx^2 + 120x = 8$, s is a constant. Which of the following represents the value of x?

A) $\frac{-120 \pm \sqrt{(120)^2 - 32s}}{2s}$

B) $\frac{-120 \pm \sqrt{(120)^2 + 32s}}{2s}$

C) $\frac{-120 \pm \sqrt{(120)^2 - 32s}}{2}$

D) $\frac{120 \pm \sqrt{(120)^2 - 32s}}{2s}$

3.

Week	1	2	3	4	5
Northridge	10.1	12.2	14.1	11	8.7
Balonton	4	7.1	6.4	11.2	15

The given table shows the average temperatures in 2 cities during 5 consecutive weeks in degrees Celsius. Which of the following is true?

A) The standard deviation in average temperatures during the 5 weeks was greater in Northridge than in Balonton.
B) The standard deviation in average temperatures during the 5 weeks was greater in Balonton than in Northridge.
C) The standard deviation in average temperatures during the 5 weeks was the same in both cities.
D) There is not enough information to compare the standard deviations in average temperatures during the 5 weeks for these 2 cities.

4.
$$\frac{5z - 1}{2} = 6az$$

In the given equation, a is a constant. What value of a would result in an equation with no solution?

5.

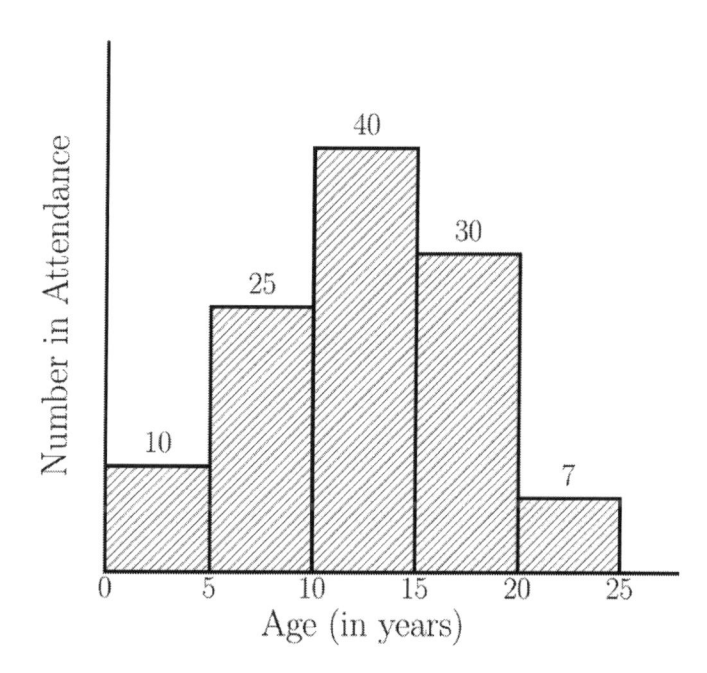

The histogram shown represents the age of the people who attended a school concert who were under the age of 25. In this figure, the 1st bar represents ages greater than or equal to 0 but less than 5. The next bar represents ages greater than or equal to 5 but less than 10, and so on. Which of the following could be the median age of these people?

A) 9
B) 10
C) 14
D) 21

6. Based on the histogram above, what is the smallest possible range of the age of the people who attended who were under the age of 25? (Consider all ages to be integer values.)

7.

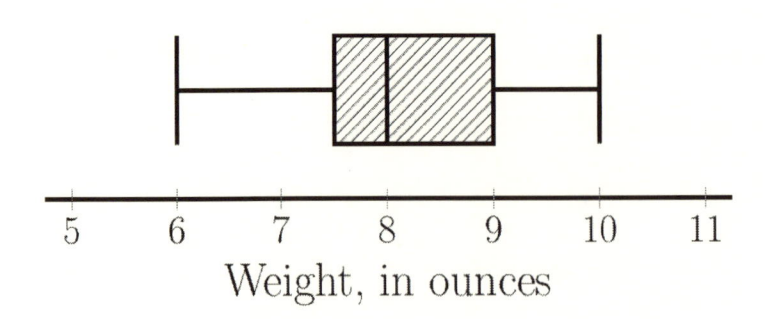

The box plot shown above represents the weight in ounces of 8 toads. If the median weight of the toads is 8 ounces, what is the maximum number of these toads that could weigh exactly 8 ounces?

A) 0
B) 1
C) 2
D) 4

8.

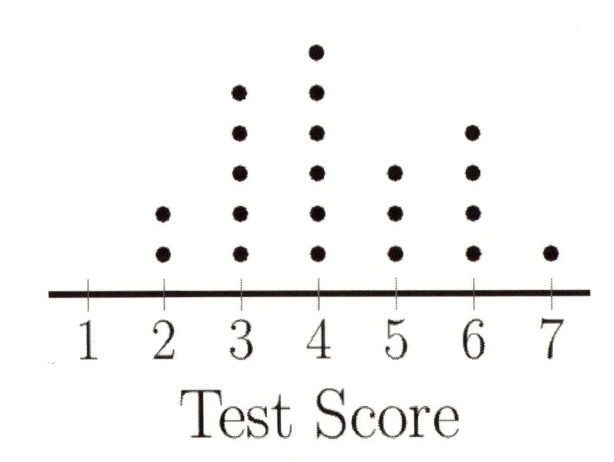

The dot plot shown displays the 21 test scores given to a class. The possible scores are the integers from 1 to 7. If 2 scores are chosen at random, what is the probability that the 1st score chosen is a 3, and the next score chosen is a 6?

9.
$$\frac{r+s+t}{r} = \frac{2r+t}{s}$$

In the equation above, r and s are constant non-0 values. There would be infinite solutions for t if which of the following were true?

A) $r = 1$
B) $r = 0$
C) $r = s$
D) $r = 2s$

10.

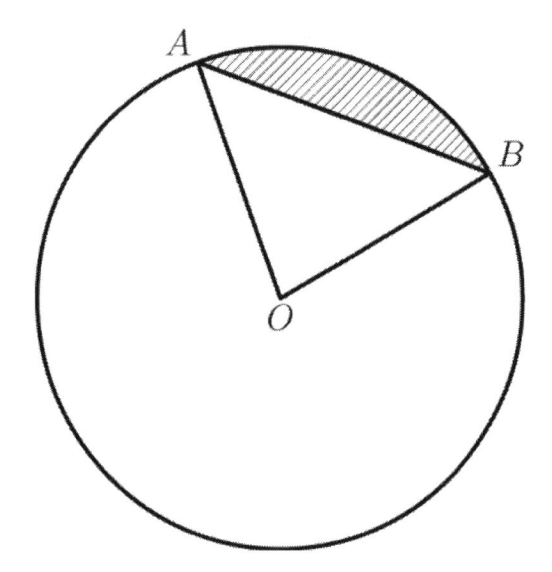

In the figure above, points A and B lie on the circumference of circle O, and the measure of angle OBA is 60 degrees. If the radius of the circle is 5, what is the area of the shaded region?

A) $\frac{50\pi - 75\sqrt{3}}{12}$
B) $\frac{50\pi - 75\sqrt{3}}{12}$
C) $\frac{50 - 75\sqrt{3}}{2}$
D) $\frac{50\pi - 75\sqrt{2}}{2}$

11. $x^4 - rx^3 - sx^2 - 8x = (x^2 + 2x)(x^2 - 4)$

In the equation above, r and s are constants. If this equation is true for all values of x, what is the value of rs?

12. Alex and Celia who live 1,200 miles apart will drive nonstop toward each other to meet. If Alex leaves his home at 10:00 a.m. and drives at an average speed of 60 miles per hour, and if Celia leaves her home the same day at 8:00 a.m. and drives at an average speed of 50 miles per hour, how long will Celia be driving until they meet?

A) 8 hours
B) 9 hours
C) 10 hours
D) 12 hours

13. In an arithmetic sequence, $n_3 = -12$, and $n_5 = 6$. What is the value of n_{30}?

14. In the xy-plane, a triangle has vertices at $(-4, 9), (-4, 2)$, and $(10, 2)$. What are the coordinates of the triangle after it has been translated down 2 units and then reflected over the x-axis?

Quiz 3 | 219

A) $(-4,-7), (-4,0)$, and $(8,0)$
B) $(-4,-7), (4,-2)$, and $(8,-2)$
C) $(-4,-7), (-4,0)$, and $(10,0)$
D) $(4,-7), (4,0)$, and $(-10,0)$

15. If $\sin(x + 1)° = \cos(y - 5)°$, what is the sum of x and y?

16. The length of the hypotenuse of triangle ABC is 8, and the length of leg AB is $4\sqrt{2}$. What is the measure of angle ACB in radians?

A) $\frac{\pi}{8}$
B) $\frac{\pi}{4}$
C) $\frac{\pi}{2}$
D) 45π

17.
$$-5 < x < 11$$

Which of the following is equivalent to the inequality shown above?

A) $|x - 3| < 8$
B) $|x - 5| > 8$
C) $|x - 8| < 3$
D) $|x - 3| < 16$

18.
$$\frac{x+y}{y^2} - \frac{y}{y^2-1} = \frac{1}{y^2-1}$$

Given the equation shown, which of the following is the solution for x in terms of y?

A) $x = -\frac{y}{y-1}$
B) $x = \frac{y}{y^2-1}$
C) $x = \frac{y}{y-1}$
D) $x = \frac{y+1}{y}$

19. The equation $\sqrt{a + 12} = a$ has how many real solutions?

A) 0
B) 1
C) 2
D) Infinite

20. An economist, after first determining the number of stores that sell eggs in the city of Benvale on one particular day, is then able to find out all the prices of eggs at about 85% of these stores on that day. If the economist were to then approximate from this evidence the cost of eggs sold at stores throughout the city of Benvale that day, the answer to which of the following questions would be most helpful in determining the accuracy of their conclusion?

A) Which of the stores for which the economist found prices sold the most eggs?
B) How much does the price of eggs sold in Benvale fluctuate?
C) What percentage of the eggs sold in Benvale that day were sold at the stores for which the economist was able to find prices?
D) Are most of the eggs sold at the stores for which the economist was able to find prices bought by residents of Benvale?

Day 20

Rest, Review, Quiz 3 Explanations

1.
$$\frac{\frac{5}{x^2-3} + \frac{y}{2x}}{\frac{1}{2}}$$

Which of the following is equivalent to the fraction given above?

A) $\frac{5x+x^2y-3y}{x^2-3x}$

B) $\frac{5x+2x^2y-6y}{x^3-3x}$

C) $\frac{10x+x^2y-3y}{x^3-x}$

D) $\frac{10x+x^2y-3y}{x^3-3x}$

With a complicated fraction like this, let's first simplify the numerator. You're adding 2 fractions, so you'll need a common denominator. We'll get that by multiplying $x^2 - 3$ and $2x$, giving us $2x^3 - 6x$. That gives us $\frac{\frac{10x+x^2y-3y}{2x^3-6x}}{\frac{1}{2}}$. To divide by ½, we'll multiply the top by 2, giving us $\frac{10x+x^2y-3y}{2x^3-6x} \times 2$. One efficient way to proceed from here is to **factor** that 2 from the terms in the denominator, giving us $\frac{10x+x^2y-3y}{2(x^3-3x)} \times 2$. We can now cancel both 2's so we have $\frac{10x+x^2y-3y}{x^3-3x}$, which is **choice D**.

2. In the equation $sx^2 + 120x = 8$, s is a constant. Which of the following represents the value of x?

A) $\frac{-120 \pm \sqrt{(120)^2 - 32s}}{2s}$

B) $\frac{-120 \pm \sqrt{(120)^2 + 32s}}{2s}$

C) $\frac{-120 \pm \sqrt{(120)^2 - 32s}}{2}$

D) $\frac{120 \pm \sqrt{(120)^2 - 32s}}{2s}$

It's clear from the choices that we'll be using the quadratic formula for this one. Once we subtract 8 from both sides, we can see that $a = s$, $b = 120$, and $c = -8$. Before doing much else, **consult the choices**. Knowing the quadratic formula, we can rule out choice D because it doesn't begin with **negative** 120. Also, you should notice that choice C has 2 in the denominator instead of $2s$, so we can

222 |Day 20

rule that out. Since the discriminant (the part under the root sign) should be $b^2 - 4ac$, we want $120^2 + 32s$, as it shows in **choice B**.

3.

Week	1	2	3	4	5
Northridge	10.1	12.2	14.1	11	8.7
Balonton	4	7.1	6.4	11.2	15

The given table shows the average temperatures in 2 cities during 5 consecutive weeks in degrees Celsius. Which of the following is true?

A) The standard deviation in average temperatures during the 5 weeks was greater in Northridge than in Balonton.
B) The standard deviation in average temperatures during the 5 weeks was greater in Balonton than in Northridge.
C) The standard deviation in average temperatures during the 5 weeks was the same in both cities.
D) There is not enough information to compare the standard deviations in average temperatures during the 5 weeks for these 2 cities.

Standard deviation is all about how spread out the data is from the average, but we won't bother to calculate the averages; we'll approximate. For Northridge, it is somewhere around 11. For Balonton, the average is closer to 8. For Northridge, none of the numbers are very far from 11, but in Balonton, some of the numbers such as 4 and 15 **are** "pretty far" from 8, so this set of data has a greater standard deviation, which means the answer is **choice B**. Note: If you have trouble estimating the average, just add the integer values, and divide by 5.

4.
$$\frac{5z - 1}{2} = 6az$$

In the given equation, a is a constant. What value of a would result in an equation with no solution?

With an equation like this, let's "flatten it out" by multiplying each side by 2, giving us $5z - 1 = 12az$. We know this will have no solution if the coefficients of z are the same, since 1 side has a minus 1, and the other does not. Therefore, if a is equal to $\frac{5}{12}$, we'll have $5z - 1 = 12(\frac{5}{12})z$ or simply $5z - 1 = 5z$. This equation has no solution.

Rest, Review, Quiz 3 Explanations | 223

5.

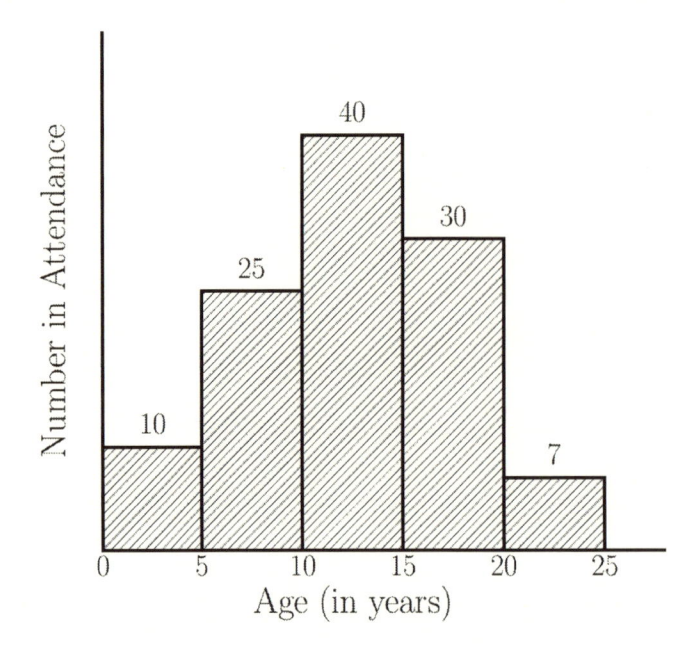

[Histogram: Number in Attendance vs Age (in years). Bars: 0–5: 10; 5–10: 25; 10–15: 40; 15–20: 30; 20–25: 7.]

The histogram shown represents the age of the people who attended a school concert who were under the age of 25. In this figure, the 1st bar represents ages greater than or equal to 0 but less than 5. The next bar represents ages greater than or equal to 5, but less than 10, and so on. Which of the following could be the median age of these people?

A) 9
B) 14
C) 15
D) 16

If we add up the number of people in attendance, we get 112. That means the median value of these ages will be the mean (average) of the 56th and 57th age if they were all listed in ascending (or descending) order. The first 35 ages would all be at least 0 and less than 10. The 56th and 57th ages will be in the next category, which has the next 40 ages. The average of these 2 ages will also be in this category, which means the only possible median of these choices would be 14, which is **choice B**.

6. Based on the histogram above, what is the smallest possible range of the age of the people who attended who were under the age of 25? (Consider all ages to be integer values.)

To get the smallest range, we'll choose the largest minimum and smallest maximum values. It could be that the 1st bar represents only people who are 4. The last bar could include only people who are 20. That makes the range 16.

7.

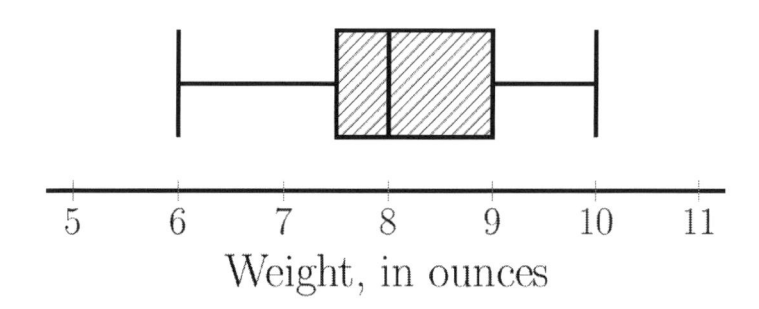

![Box plot showing weights in ounces from about 6 to 10.5, with the box from about 7 to 9 and median at 8.]

The box plot shown above represents the weight in ounces of 8 toads. If the median weight of the toads is 8 ounces, what is the maximum number of these toads that could weigh exactly 8 ounces?

A) 0
B) 1
C) 2
D) 4

Of the 8 toads, we know we have at least 1 that is 6 ounces, or at least quite close to 6. Likewise, we have at least 1 toad that is 10 ounces, or very near 10. Let's look at the choices. Since the question asks for the maximum possible number of 8-ounce toads, could it be the largest choice given, which is 4? The answer is yes. For example, the toads could have the following weight in ounces: 6, 7, 8, 8, 8, 8, 10, 10. The median is 8, as it looks in the figure. The median of the 4 values "on the left" (6, 7, 8, 8) is 7.5, as it looks. The median of the other 4 (8, 8, 10, 10) is 9, as it looks. **Choice D is correct.**

8.

![Dot plot of test scores from 1 to 7.]

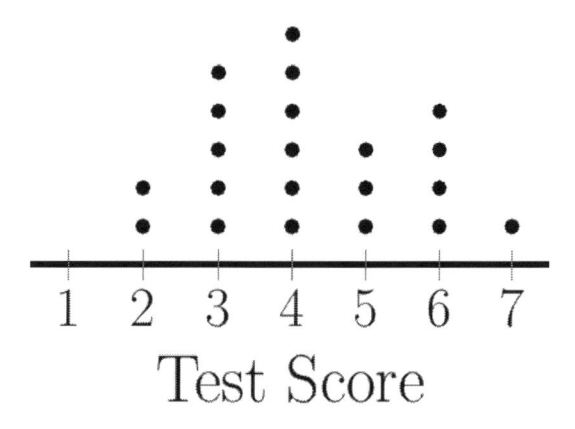

The dot plot shown displays the 21 test scores given to a class. The possible scores are the integers from 1 to 7. If 2 scores are chosen at random, what is the probability that the 1st score chosen is a 3 and the next score chosen is a 6?

The probability of randomly choosing a 3 is $\frac{5}{21}$ because there are 21 possibilities, 5 of which are 3. The probability of randomly choosing a 6 is $\frac{4}{20}$ because there are now only 20 possibilities, 4 of which are 6. To get the probability that both events occurred, we multiply $\frac{5}{21} \times \frac{4}{20}$. You can do some cancellation, and you'll get $\frac{1}{21}$. This can also be entered as .0476.

Rest, Review, Quiz 3 Explanations | 225

9.
$$\frac{r+s+t}{r} = \frac{2r+t}{s}$$

In the equation above, *r* and *s* are constant non-0 values. There would be infinite solutions for *t* if which of the following were true?

A) $r = 1$
B) $r = 2$
C) $r = s$
D) $r = 2s$

This one is almost certainly best answered by trying out the choices. Once we try **choice C**, we can substitute *s* for *r*, giving us $\frac{2s+t}{s} = \frac{2s+t}{s}$. We could go further, but it might already be apparent to you that *t* can be any number. If we did simplify further, we would end up with $t = t$. That, of course, is true for any value.

10.

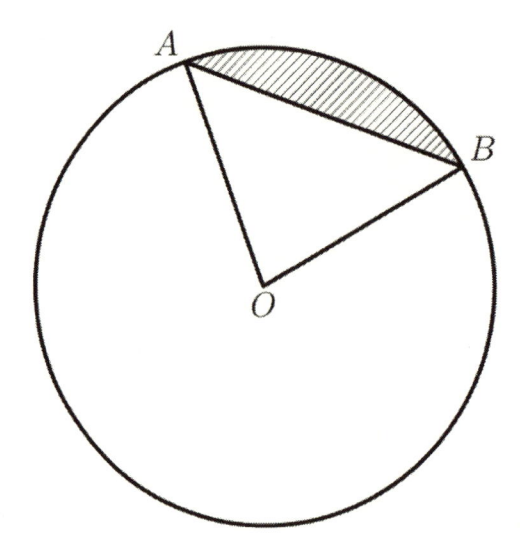

In the figure above, points *A* and *B* lie on the circumference of circle *O*, and the measure of angle *OBA* is 60 degrees. If the radius of the circle is 5, what is the area of the shaded region?

A) $\frac{50\pi - 75\sqrt{2}}{12}$
B) $\frac{50\pi - 75\sqrt{3}}{12}$
C) $\frac{50 - 75\sqrt{3}}{2}$
D) $\frac{50\pi - 75\sqrt{2}}{2}$

The area of the shaded region is equal to the area of that sector (slice) of the circle minus the area of the triangle. Since we know the radius is 5, we can say that the area of the circle is 25π. We can also conclude that this is an equilateral triangle (every side is 5) since the radii are (by definition) equal and

226 |Day 20

we're given angle *OBA*'s measure as 60 degrees. Since this is 1/6 of 360, we know that the area of the sector is 1/6 of 25π, or $\frac{25\pi}{6}$.

You can split this triangle into 2 30/60 right triangles by drawing its height (altitude) from the center to the point on the triangle between *A* and *B*. We know that the sides of a 30/60 are in the ratio 1 : $\sqrt{3}$: 2. Since our base (*AB*) is **5,** the short leg is $\frac{5}{2}$, and the long leg, *which is the height*, is $\frac{5}{2} \times \sqrt{3}$ or $\frac{5\sqrt{3}}{2}$. That means the area of the triangle (base x height divided by 2) is $\frac{(5)(\frac{5\sqrt{3}}{2})}{2}$. Dividing by 2 is like multiplying by ½, so you get $\frac{25\sqrt{3}}{4}$.

The area of the shaded region is therefore $\frac{25\pi}{6} - \frac{25\sqrt{3}}{4}$. With a common denominator of 12, we'll get **choice B**.

That took a while. Are there any shortcuts, please? Well, it shouldn't take long to realize that the area of the sector is going to have π in it. Circles and parts of circles almost always will. That would at least eliminate choice C. Once you realize you're dealing with a 30/60, you might recognize that this almost always means we'll have $\sqrt{3}$ somewhere in the area of the triangle, and that would knock off choices A and D.

11.
$$x^4 - rx^3 - sx^2 - 8x = (x^2 + 2x)(x^2 - 4)$$

In the equation above, *r* and *s* are constants. If this equation is true for all values of *x*, what is the value of *rs*?

This should look like a **matching coefficient** question. The telltale phrase is "for all values of *x*." Let's get started by distributing (FOIL-ing) the right-hand side, giving us the following:

$$x^4 - rx^3 - sx^2 - 8x = x^4 + 2x^3 - 4x^2 - 8x$$

The *x* values that are raised to the 4th power already have matching coefficients of 1, so we don't have to worry about them. If you want, you can subtract them from both sides. We also have $-8x$ on both sides, so these can be ignored as well. The coefficient of the *x* values raised to the 3rd power are negative *r* on one side and 2 on the other side. These need to match, so we know that *r* is equal to **negative 2**. Also, the coefficients of the *x* values raised to the 2nd power are negative *s* on one side and negative 4 on the other, so we know that **s = 4**. Therefore, the value of *rs* is $(-2)(4) = -8$.

12. Alex and Celia who live 1,200 miles apart will drive nonstop toward each other to meet. If Alex leaves his home at 10:00 a.m. and drives at an average speed of 60 miles per hour, and if Celia leaves her home the same day at 8:00 a.m. and drives at an average speed of 50 miles per hour, how long will Celia be driving until they meet?

A) 8 hours
B) 9 hours
C) 10 hours
D) 12 hours

This question clearly involves the use of the formula Rate x Time = Distance. We can call Alex's time t, which means we have to call Celia's time $t + 2$ because she will have been driving for 2 hours more when they meet, having left 2 hours earlier. Since Alex's rate was 60 mph, we can express his distance as $60t$. Celia's distance will be $50(t + 2)$. Note that each of these represents each driver's distance **when they meet** somewhere between their 2 starting points. Since they started out 1,200 miles apart, we can say $60t + 50(t + 2) = 1200$. In other words, when they meet, the **sum** of their distances will have been 1,200. This is a fairly simple equation, which gives us a solution of $t = 10$. Let's be careful, however. The question asks for Celia's time, which is $t + 2$, or 12 hours. As it turns out, that means they would meet exactly in the middle, but this is not implied by the structure of the question. At different speeds or starting times, that would not be the case.

13. In an arithmetic sequence, $n_3 = -12$ and $n_5 = 6$. What is the value of n_{30}?

From the 3rd number to the 5th number, 18 is added. That means 9 is added between consecutive values in the sequence. You could determine the 1st number in the sequence, or you could, for example, determine the answer starting with $n_5 = 6$. From the 5th term to the 30th term, we must add 9 a total of 25 times. Since 9 times 25 is 225, we add 225 to 6 to get 231.

If you went backward to the 1st term, that would give you -30. You would then add 9 a total of 29 times to this number, which would, of course, also still give you 231.

14. In the xy-plane, a triangle has vertices at $(-4,9), (-4,2)$, and $(10,2)$. What are the coordinates of the triangle after it has been translated down 2 units and then reflected over the x-axis?

A) $(-4,-7), (-4,0)$, and $(8,0)$
B) $(-4,-7), (4,-2)$, and $(8,-2)$
C) $(-4,-7), (-4,0)$, and $(10,0)$
D) $(4,-7), (4,0)$, and $(-10,0)$

When translating down 2, all the y values decrease by 2 while the x values remain unchanged. That gives us $(-4,7), (-4,0)$, and $(10,0)$. If we now flip (reflect) this over the x-axis, only the y values change sign. A value of 0 will remain 0. That gives us **choice C**.

15. If $\sin(x + 1)° = \cos(y - 5)°$, what is the sum of x and y?

This question hinges on the fact that if 2 angles sum to 90 degrees, then the sine of 1 of them equals the cosine of the other, in both directions. Here's another way to say this: "If the sine of 1 angle equals the cosine of another, they sum to 90 degrees." Since the sine of $x + 1$ degrees equals the cosine of $y - 5$ degrees, it must be true that $(x + 1) + (y - 5) = 90$. We can simplify this to $x + y - 4 = 90$ and $x + y = 94$.

16. The length of the hypotenuse of right triangle *ABC* is 8, and the length of leg *AB* is $4\sqrt{2}$. What is the measure of angle *ACB* in radians?

A) $\frac{\pi}{8}$
B) $\frac{\pi}{4}$
C) $\frac{\pi}{2}$
D) 45π

You might recognize that this is a 45/45 triangle with sides in the ratio $1 : 1 : \sqrt{2}$. It's a little tricky to recognize it here since we find the root sign in the leg, not the hypotenuse. However, if you multiply the leg by root 2, you will indeed get 8. But if you didn't recognize it, you can still use the Pythagorean theorem to determine that the other leg is also $4\sqrt{2}$.

Whatever method you use, you'll get a triangle like this one.

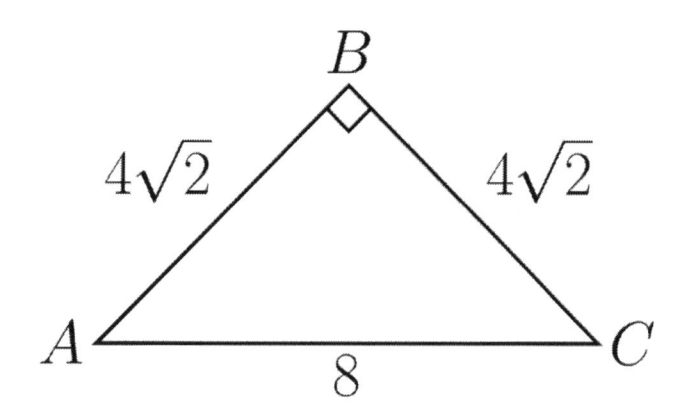

The angle we need is 1 of the 45 degree angles, and (remember, 180 degrees equals π radians) that is equal to $\frac{\pi}{4}$ radians, which is **choice B**.

17.
$$-5 < x < 11$$

Which of the following is equivalent to the inequality shown above?

A) $|x - 3| < 8$
B) $|x - 5| > 8$
C) $|x - 8| < 3$
D) $|x - 3| < 16$

You might best tackle this by turning the answer choices into inequalities. When we get to **choice A**, we can say that $x - 3 < 8$, **and** $x - 3 > -8$. Simplified, we get $x < 11$ and $x > -5$. Together, that means $-5 < x < 11$.

Rest, Review, Quiz 3 Explanations | 229

18.
$$\frac{x+y}{y^2} - \frac{y}{y^2-1} = \frac{1}{y^2-1}$$

Given the equation shown, which of the following is the solution for *x* in terms of *y*?

A) $x = -\frac{y}{y-1}$
B) $x = \frac{y}{y^2-1}$
C) $x = \frac{y}{y-1}$
D) $x = \frac{y+1}{y}$

The 1st thing to notice is that 2 of the fractions have a common denominator. It makes sense to get them together.
$$\frac{x+y}{y^2} = \frac{y}{y^2-1} + \frac{1}{y^2-1}$$

Now we can subtract, giving us this:
$$\frac{x+y}{y^2} = \frac{y+1}{y^2-1}$$

The next thing to recognize is that $y^2 - 1$ is *the difference of 2 squares* and can be factored into $(y+1)(y-1)$. This is great because we can now cancel the $y+1$ on the top and bottom of this fraction, giving us $\frac{x+y}{y^2} = \frac{1}{y-1}$. Our goal is to get *x* by itself, so let's do what we usually do in this situation and "flatten" the fraction by cross-multiplying. That gives us $y^2 = xy - x + y^2 - y$. If we subtract y^2 from each side and, while we're at it, add *y* to each side, we get $y = xy - x$. What's next? *When in doubt, factor out.* We can factor the *x* from the terms on the right, giving us $y = x(y-1)$. We do that so we can now isolate *x* by dividing each side by $y - 1$, giving us **choice C**.

Are there any shortcuts? Well, you could graph the original equation and then graph each choice until 1 matches it exactly. If you use the "copy" feature on the OSC, this won't take too long.

19. The equation $\sqrt{a+12} = a$ has how many real solutions?

A) 0
B) 1
C) 2
D) Infinite

Let's get rid of that root sign by squaring each side, giving us $a + 12 = a^2$. This is looking like a quadratic, so we can re-form this as $a^2 - a - 12 = 0$. This factors into $(a-4)(a+3) = 0$ and has 2 solutions: 4 and -3.

However, when you're dealing with square roots, you should always check your answers. If we try 4 in the original equation, there's no problem. But if we plug in -3 for *a*, we'll get the false statement $3 = -3$. Therefore, the only solution is 4, leading us to **choice B**.

20. An economist, after first determining the number of stores that sell eggs in the city of Benvale on one particular day, is then able to find the prices of eggs at about 85% of these stores on that day. If the economist were to then approximate from this evidence the cost of eggs sold at stores throughout the city of Benvale that day, the answer to which of the following questions would be most helpful in determining the accuracy of their conclusion?

A) Which of the stores for which the economist found prices sold the most eggs?
B) How much does the price of eggs sold in Benvale fluctuate?
C) What percentage of the eggs sold in Benvale that day were sold at the stores for which the economist was able to find prices?
D) Are most of the eggs sold at the stores for which the economist was able to find prices bought by residents of Benvale?

The conclusion is about eggs sold in Benvale. Our sample should be as large and as representative as possible. The only choice that addresses either of these factors is choice C. Let's say there are 100 egg-selling stores in Benvale, so the economist got data from 85 of them. But what if the other 15 stores were huge supermarkets that were responsible for the sale of 90% of eggs sold in Benvale? That would weaken any conclusion the economist would draw.

Day 21

Test Experience, Reading and Writing Section, Reviewing Test 1

After reading this material and reviewing Test 1, sign and check the box for Day 21.

The next 3 days are for gearing up for the final push, preparing for the test experience, checking any admission boxes that haven't been checked, and yes, **reviewing.** Remember, it's not simply the students who know the most who do the best. It's the students who know the most and who **use** what they know. And no one gets to that point without integrating their knowledge so they don't waste time thinking, "Um, what do I do with a negative exponent? How do you complete the square with a quadratic?" This full integration comes from repetition and then more repetition, and from a relentless pursuit of gaps in your knowledge. Our aim here is the confidence that comes from mastering the material. Now's the time!

A good place to start is with **a review of Test 1** that you took on Day 1. Have another look at it. Notice how far you've come. But you'll almost certainly see some questions that are still just a little sticky. That's great! Focus on them like a laser beam. Next time you see a question of that sort, smile to yourself, knowing that you'll be able to handle it with confidence and efficiency.

Note: For explanations of all the practice tests, we recommend Khan Academy rather than the official explanations found with the test. (You'll see a link to Khan on the official test.)

It might be that when you first did this test, you were given an easier Module 2A. Why not try Module 1 again and see if you can improve your performance to the point that you get the tougher (and worth more to your score) Module 2B? Of course, afterward, study the explanations for any questions that gave you a hard time, **even if you got them right.**

Let's talk about **the test experience**.

By now, you should know exactly where you need to be and when. You know to bring your photo ID, a snack, pencils, and, of course, the fully charged device with the downloaded Bluebook app on which you'll be taking the test (unless you're using a device provided by the school or other test-taking facility). If it's your device, sometime between 1 and 5 days before the exam date, you'll do a quick exam setup that will generate an admission ticket. You can also bring an approved calculator (not one that makes sounds or prints), although you'll almost certainly rely on the OSC.

Backup devices are allowed, but only 1 can be open and available at any time.

If you're taking the test **not** in your home country, you'll almost certainly need a passport as your ID. If you're unsure, check with the testing facility beforehand.

Once you are seated (likely in an assigned seat), you'll get connected to the Wi-Fi network and log on to Bluebook. The proctor will give you a start code, and once you enter it, away you go. Not everyone will be starting or completing sessions at exactly the same time, so ignore all the other test-takers. You can raise your hand to get the proctor's attention if you have an issue (of course, not a test content issue).

If you have a phone with you, it will be collected or placed somewhere beyond your reach. Make sure it's off. If it makes any sound, they could dismiss you and cancel your score. You cannot access the phone during the break.

First up is Reading and Writing (R&W). (That's 1 section, not 2.) As with Math, you'll get 2 modules of R&W, and the 2nd one is dependent on the 1st. Each module has 27 questions to complete in 32 minutes. You can skip around within a module and mark a question to go back to it later. You cannot go back to the previous module, however.

For the R&W modules, you can highlight part of a passage or make notes. You can zoom in and out. As you might know, the passages are much shorter than the ones on the older paper tests. They are typically 1 paragraph or 4 to 6 bullet points, or there may be 2 paragraphs you need to compare in some way. Graphic material such as tables will appear. All questions are multiple choice, and there are no penalties for incorrect answers, so always make a choice.

After the 2 R&W modules, you'll get a 10-minute break when you can get up and stretch. Make sure you take your ID with you if you leave the room. Stretching is no little thing. Keeping the body relaxed and getting oxygen to the brain is crucial for optimum performance. It's fine to have a small snack or some water. If you do have to take a break during the test-taking time, you will not get extra time.

If you have any questions about an administrative or technical issue, a counselor at your school should be able to help.

If you haven't already, take another look at Test 1. Give yourself some back-pats, and then find some remaining gaps in your test knowledge. This next week is where you'll literally put that knowledge to the test!

Day 22

The Admissions Process

After reading this material, sign and check the box for Day 22.

With all the focus on the exam, it's important not to lose sight of the fact that it's only 1 part of the admissions process, not the whole thing. Today is a day for checking (not literally) other boxes that have nothing to do with the test.

The Recommendation Letter

Have you thought about who will be writing a recommendation letter for you? Ideally, there's a teacher who knows you well and can say glowing things about you. They don't have to say that you're brilliant, just that you're motivated, curious, and an independent thinker but also a terrific team player. The admissions people at colleges read many of these, so anything that might make them pause and think, "hmmm . . . interesting" (well, *almost* anything) is what you want. And if it seems like the teacher **really does know you** and that you made an *impression* on them and has a great anecdote about that—great!

You will probably also want a letter from a non-teacher—adults you've worked with either for pay or as a volunteer, or a coach. A well-known person is **not** a good idea unless it's apparent that you really have a relationship with this person.

The people you ask are likely to be busy folks, and high school teachers get many requests of this sort. If you really want to get a letter from someone in particular, ask early, and ask often.

The Application Essay (Personal Statement)

Depending on where you're applying, you'll need to write a certain number of essays on a variety of topics. If you use the Common App, you might end up writing only 1 essay that will then be sent to all the schools that accept it. Not all schools do, however, and some will require more.

These essays are sometimes called personal statements—and for a reason. The school can get a pretty good sense of your academic performance and potential from your school records and test scores. They might even get some sense of who you are from the recommendation letters and your non-academic resume. But they want more, especially the highly selective schools that get so many applications from high-achieving students.

A good application essay will provide information about you that is, for lack of a better word, intimate. The admissions people want, as much as they can, to get to know you and picture you.

Remember, all good writing forms pictures in our minds. You can do that with specifics. For example, you *could* write the following:

> *My own background has made me sensitive to how difficult life can be for the elderly. When you can't get around easily anymore, when your family isn't around, when your income is fixed, that can be so difficult for an individual. That's why I'm interested in pursuing my academic interest in geriatric law.*

That's not bad, but what about this:

> *One day I was walking up our driveway when I saw a woman in a bathrobe leave the house a few doors down and start walking along the sidewalk. I realized that this must be Ms. Pattin. I'd only seen her once, before she was diagnosed with dementia. I approached her.*

You might say, "Sure, but I can only say that *if it happened*. And yes, of course that's true. The point, however, is that a *story* beats a *description* every time. So maybe you didn't rescue Ms. Pattin, but what about this?

> *I've pictured what life will be like for my parents one day, and the truth is that I'm a little nervous. They're healthy now, but one day they might have to move to a facility that can take care of them. I know from what I've seen at my grandfather's residence (which isn't bad, just sad) and from horror stories I've read and heard from others (not enough nurses, unclean facilities, exorbitant rates) that there is no guarantee that these wonderful, giving people who raised me will live the life they deserve. The truth is that this makes me mad, but getting mad is pointless—selfish really. That's why . . .*

You want the people who read your essay to have a little movie in their minds so they're **not** thinking, "Here I am at 11:40 p.m. reading yet another essay." Instead, you want them to think, "Then what?" or "Oh, that was a bit of a surprise." That may happen **if you can really pull it off**.

Now, be sure to **edit** your essay.

Editing is very important, so we'll say it again.

Edit.

After you've edited your writing a number of times, **crafting** a polished, concise, intriguing essay, give it to someone else to look at—perhaps 2 people whose opinions you respect. Then, after you've made some changes, put it aside for a few days or a week or so. Then—

Edit it again, and proofread it.

School Selection

School selection is, of course, a huge topic that we cannot discuss in any length here. But here's what we will say.

These days, there are many ways to investigate what schools you'd like to attend. Perhaps you can visit them and have a guided tour. If not, many schools have virtual tours online. You can ask for promotional materials, but be a little skeptical. They are, after all, advertisements. With the school's help, you can perhaps get the help of a guidance counselor and talk to current or former students. There are also a number of books and online resources that will summarize data about colleges, including how people feel about their experiences at the institution.

You'll want to pay attention to how selective a school is, what kinds of resources they have (labs, gyms, piano lessons, etc.), and what their policy is regarding early admission and early decision. Someone at your school can help clarify those terms. Of course, you'll also want to know the size of the school and its location.

Some schools say they're *test-optional*, but let's think about what that means and how it plays out in the admissions process. Admissions officers have 3 piles: accept, decline, and maybe. They are looking at reasons to move the process along and put *your* application into one of those piles. Let's say you have an excellent GPA, a terrific SAT score, and you've worked in an animal clinic for 2 years. But so has another applicant, **except they've chosen not** to submit a test score.

Now picture yourself as the admissions officer. What do you think you'd do? In one case, there's an unknown factor, and in the other case, there's a **known, highly favorable factor**—that terrific test score. Hmm.

Extracurricular Activities

The clichés you've perhaps heard about extracurricular activities are true. Colleges want to see that you're a person with strong and even passionate interests in something—perhaps a sport or a volunteer activity. Volunteer work is pretty much expected these days. And yes, *leadership* is highly valued. If you've ever started a club, organized an event, started a band, or done something else, tell the college about it. If you had to abandon your comfort zone to do so, tell them about that too. And if you just love leading, teaching, or organizing a bunch of people, be humble, but tell them that too.

You might think that at this stage there's not much you can do to add something impressive to your application, but there is no shortage of volunteer activities out there. Can you find one that really moves you? Are you interested in politics? Perhaps there's something you can do at a local elected official's office. Take it one step further. Can you bring along 2 friends and actually approach the official with an idea for some activity, maybe one that involves your school? Get creative.

GPA

Of course, your GPA is a crucial part of your admissions package. How'd you do, and what courses did you take? If you need to rescue a drooping physics grade, get on it right now! The schools love students who have taken challenging courses and met the challenge.

If you're a senior, you only have so much time to polish your GPA and firm up your application with some awesome accomplishments. But we have good news. At this stage, here's the best way to improve your chance of being admitted to the college you want to attend.

Master the test! (You're almost there.)

Day 23

The OSC in Depth, Mental Math, Time-Saving Tips, Nerves

After reading this material, sign and check the box for Day 23.

You're about to complete 3 full-length digital SAT exams. You'll get to apply what you know and find those areas that still need work. Today, we'll look in more depth at the OSC, discuss mental math shortcuts and time-saving tips, and learn how to handle nerves.

Getting the Most out of the OSC

We've seen how useful the **OSC** can be for a range of questions. Let's take a closer look at its features and how they can help you handle questions—including some of the toughest ones—in the most efficient way possible. Some of these features are not obvious to the casual user, so it's important to know them all and practice with the real thing. That can make a big difference in your confidence and your timing—in other words, to your score!

You can enter an algebraic expression or equation in whatever form you want. You can enter a quadratic in vertex form or otherwise, and the graph of the parabola will look the same. Or you can enter the equation of a line with, for example, x isolated on 1 side rather than as $y = mx + b$.

You can use the on-screen keyboard to easily notate exponents, roots, fractions, and so forth. There's a button that automatically squares a value. You can then change the "2" to "3" if you want to cube it instead. Or you could use the a^b button. For fractions, you can use the division sign on the on-screen keyboard or the slash mark on your computer keyboard. If you have a decimal that you want to view as a fraction, you can click on the little fraction symbol directly to the left of the decimal.

You can see a helpful list of x and y values for your graph. Whatever the equation you've entered, the *gear* icon will allow you to choose to see a list of values that fit your graph. Once you see the list, you can enter an x value under that column, and just like that, you'll see the correct y value. Try it!

You can start with a list of x and y values and then "connect the dots." The + sign on the upper left allows you to open a table. You can then quickly enter a bunch of x and y coordinates that will appear as points. If they're not visible, click on the magnifying glass, and the graph will center the points. If you click and hold on the round "buttonhole" at the top of the y column, you can select "line" to connect the points. (You can also select "drag" if you want to move them, although you probably won't need to do that.)

You can use function notation, and it remembers your function. Let's say you have a question that gives you $f(x) = x^2 - 2x + 3$ and then tells you $g(x) = 2f(x - 1) + 8$. If you enter the 1st equation and underneath it enter the 2nd equation, the calculator knows that the 2nd equation is a modification of the 1st one. Note: That won't happen if you use y instead of $f(x)$.

You can use it for trig, although you might not always want to. Let's say you need to know the cosine of 30 degrees. First, you have to make sure the OSC is set up for degrees, not radians. It defaults to radians, so you'll need to **click on the wrench icon and then choose degrees**. Then you can either type cos 30 or select cos after clicking *function* on the on-screen keyboard and then 30.

However, the cosine of 30 degrees is $\frac{\sqrt{3}}{2}$, but what you'll see on the screen is a long decimal because this is an irrational number. It's more likely that the test wants to know if you know that cos 30° is $\frac{\sqrt{3}}{2}$, which you can easily determine if you **really** know your 30/60 triangle, which has sides in the proportion $1 : \sqrt{3} : 2$.

You can use the slider to gradually change a value. Let's say you know the y-intercept of a line but not its slope. So you have $= mx - 1$. If you enter this into the OSC, it will say "add slider m." By clicking on the m, you can vary its value, and the slope will change as you do. It is set to show you only values from -10 to 10, but if you click on 1 of those numbers, it will allow you to enter smaller and larger numbers (in addition to the steps or increments by which it varies). There's not a great chance that you'll need to do this, but if you do want to be prepared just in case, make sure you practice.

There's a lot more the OSC can do, but it won't always save you enough time to make it worthwhile. For example, you can type "mean" and then enter a bunch of numbers, and it will find the average for you. But you can also just add them and divide by the number of terms—there's not much difference. Here, though, are a few **tips** that can help you whenever you do use the OSC.

The house icon will adjust the screen to present you with what is considered an optimum view, although it might not be the view you want. Zooming, though, is simple enough, whether you use the minus and plus sign by the "house" or use your mouse or trackpad.

Be careful about the tick marks! Just because you count up 4 ticks from the x axis doesn't mean you're at a point where $y = 4$. You should notice the scaling. Each mark might be 0.5, or it might be 100.

No commas, for example, in 1,235 will make it think you want the point with the coordinates (1, 235).

Duplicate what you've entered using the **gear** button and then the duplicate button, which looks like 2 documents.

If you want to enter the **percent sign (%)**, you should just use your computer's keypad. When you do, you'll see "% of" on the screen.

Yes, the OSC is a game-changer all right, but don't think it's the best approach for every question on the exam. With lines, parabolas, intersection points—coordinate geometry in general—it can be an enormous time-saver. And of course, for calculations you can't do on your fingers and toes—well, that's why we have calculators. And finally—

The OSC on the test is a little different from the one you're probably used to. So practice using the calculator with the official online practice material.

*

This is a good time to review the other controls you have on the digital test.

At the top of the screen is a **digital clock** so you can see how much time has elapsed. (Each Math module gives you 35 minutes to do 22 questions.) You can hide the clock, but only do that if you are completely freaked out by it or if you prefer looking at your own watch. Even then, it's best to view the on-screen clock once you have a good start on the test.

You can also hide the **OSC**, but it's unlikely that you'll want to do that. You can also move it around the screen and use the button with the 2 little arrows to change the calculator's size. There are also 2 downward-facing arrowheads next to the gear button that hide the rows where you enter information so you only see the coordinate plane.

The **mark for review** button lets you see which questions you want to return to later. You'll see a record of these questions before you submit all your answers at the end of the time for that section. This can be quite useful. You can also look at any questions you've marked this way at any time by using the question number button.

The **question number** button is at the bottom of the page. It allows you to instantly access any question or go to the review page, which shows you which questions you've marked for review and which ones you've left unanswered. Of course, you'll **never** leave a question unanswered at the end of the section. Always choose something. If you think you have no idea, are you sure you have *no* idea? You can usually eliminate at least 1 choice. Then just guess randomly.

The ~~ABC~~ **button** allows you to cross off choices you have eliminated. That can be useful, although probably more so in the Reading and Writing section.

The **reference** button at the top of the screen provides geometry formulas, 2 of the special right triangles, and a few other pieces of information. Most of this information you shouldn't need to look up. Still, some of these formulas, such as the volume of a sphere, are not typically tested, so it's nice to know you can instantly access this information if you need to.

The **more** button at the top of the screen displays a number of options such as "shortcuts" you won't need on the day of the test, as well as information about assistive technology. If you do use assistive technology such as text-to-speech, you'll have reviewed this info beforehand. The "more" button also allows you to take an "unscheduled break," but that will *not* stop the clock, so this is only for extreme cases.

And remember, you can't go on to the next module until the time is up. And you should not **want** to because it's always worth spending every second of your time making sure you haven't made a careless error.

Mental Math

Now we'll go in the opposite direction—using your brain rather than a calculator to find your way to an answer. You do not want to resort to the calculator to multiply 8 by 3. If you need to review your multiplication table, do so. That will strengthen your math brain.

And yes, you have a math brain. You might not feel that you're a natural at it, but let's make sure you fully use what you have and improve it in the process.

A lot of students get a little anxious with **decimals, fractions, and percentages**, which ultimately are all the same thing. Let's look at percentages. We mentioned multiplying 8 by 3. What about 80% of 3? You know you'll need the numbers 2 and 4. It's just a matter of where to put the decimal place. And 80% of something is "a lot of it," so we can easily say that 80% of 3 is 2.4. What about 80% of 0.3? That's 0.24. What about 8% of 0.3? That's 0.024. You'll need to fill in the decimal place with 0.

Also, since it's simple to find 10% of a number, we can easily find 20%, 30%, 40%, and so on of it. What's 30% of 110? You know that 10% of 110 is 11. Just multiply that by 3 to get 33.

And what about 15% of a number? That's just 10% of it and then another 5%. Let's say you want 15% of 280. Well, 10% of 280 is 28, and 5% would be half of that, or 14. So 15% of 280 is 28 + 14, which is 42.

That might seem like too much work or too risky since you have a beautiful calculator right there. Maybe. It's up to you, of course, on test day. But thinking that way can build those math muscles and make percentages less weird.

And remember, the test prefers fractions to the decimal or percentage equivalents. Fractions are often easier to work with. Do you want 75% of 160? Well, 75% is ¾. Since ¼ of 160 is 40, ¾ is 120.

Timesaving Tips

Tip 1

Consult the choices. Officially, each Math module (not the Reading and Writing section) gets progressively harder from the first to the last question, although by now you probably realize that the "officially" hard questions can often be solved in little time. But you should still be able to recognize when an answer choice is **too obvious to be correct**. Look, for example at the following question.

If it takes Juan 3 hours to complete a task and it takes Jules 4 hours to complete the same task, how long does it take them to complete the task working together and simultaneously?

A) 3 hours
B) 3.5 hours
C) $\frac{12}{7}$ of an hour
D) 7 hours

This is question 21 (out of 22 questions in all). It should be at least somewhat challenging, so the simple average of 3 and 4 would **not** be correct, so let's eliminate choice B. Also, choice D is wrong for a few reasons, including the fact that it's the sum of 3 and 4. You might also realize that choice A doesn't make much sense, so that leaves us with **choice C**. Of course, there's a way to solve this, which we covered on Day 18.

Sometimes consulting the choices before doing the work can save you time. Look at the following question.

A floor that is 37 feet long and 128 feet wide is being installed in a ballroom. The cost for it is d dollars per square foot plus a total labor cost of l dollars. Which of the following represents the total cost of the floor and its installation in dollars?

A) $d + l(37 \times 128)$
B) $l + d(37 \times 128)$
C) $(l + d)(37 \times 128)$
D) $ld(37 \times 128)$

If you scan the choices before doing any work, you'll see that there's no reason to calculate the area of the floor. Each choice has left that undone. This isn't an enormous time-saver, but if we can save a few seconds here and there, it's worth it. (The answer is **choice B**.)

It could also be the case that consulting the choices reminds you that an answer should be in radians, not degrees. Or maybe it should be a fraction, not a decimal. Should it be solved for y in terms of x and not the other way around?

Tip 2

Now what was that question? On a related note, before you select an answer, take another look at the question itself. If you're asked to solve $2x$, make sure you don't choose the solution for x. Are they asking you for the sum of a set of numbers or the average? What exactly do they want? For example:

$$3a - b = 42$$
$$-a + 2b = -22$$

What is the value of $2a + b$?

You *could* use substitution or elimination to solve an unknown and then the other one, and then come up with $2a + b$. *But* look at exactly what they want. It isn't necessary in this case to solve for the unknowns. If you just add the 2 equations, you'll get $2a + b = \mathbf{20}$. You're done. The question was really testing your ability to find the shortcut.

Tip 3

Automatic calculations. In some cases, you can immediately get to work once you see certain key phrases. For example, if a word problem begins by saying that the average test score for a class of 20 people is 90, you can be certain that you'll want to take these 2 parts of the average formula and find the 3rd, which would be the sum of the scores. So right away, you can determine that the sum of the scores is 20 × 90, which is 1,800. Write this down, and continue analyzing the problem. That should be an automatic reaction.

Or let's say someone drove 250 miles at an average speed of 50 miles per hour. Right away, you can conclude the time, which is distance divided by rate, or 5 hours. Continue reading the problem.

And let's not forget our special triangles. If you have a right triangle with a hypotenuse of 9 and a leg of 4.5, you know that it's a 30/60, which is the right triangle that has a hypotenuse twice the length of a leg. That means that in this case, the length of the other leg is 4.5 times root 3, which would probably look like $\frac{9\sqrt{3}}{2}$. Write that down, and proceed. This should happen **automatically**.

Tip 4

Use simple numbers. This is somewhat like the word problem strategy where you pick your own values for an unknown **when the answer choices have unknowns**. If the problem says an item that costs c dollars decreases in price by 18%, you say, "Okay, let the cost be $100." But in this case, you'll pick simple numbers just to make the problem easier to understand. Let's say you're doing a problem that has $p(1.034)^{\frac{r}{5}}$. Let's not even worry about what the problem is all about in the real world. You might be able to understand it better if you imagine that r is equal to 5. This tells you that p—whatever it is—would be multiplied by 1.034. What if r equals 10? Then p would be multiplied by 1.034 squared. What if r is 15? You get the picture. It's a way of making the problem simpler and will probably lead you to the next step.

Tip 5

Roman numeral answer choices. This might not show up on your exam, but we sometimes see cases such as the following.

Set A consists of all the odd integers greater than −10 and less than 10. Set B is the same as Set A except it also includes 1 positive even integer. Which of the following must be greater for Set B than for Set A?

I. The mean

II. The median
III. The mode

A) I only
B) I and II only
C) II and III only
D) I and III only

You might first realize that neither set has a mode, so choices C and D are gone. The remaining 2 choices both have "mean" in their answer, so it must be that the mean will increase. So all you have to consider is the median. And yes, the median of Set A is 0, and whatever the median of Set B is, it's a positive number. That means **choice B** is correct.

Tip 6

When in doubt, factor out. Students often forget that when handling an algebraic equation, you don't always have to "do something" to both sides. It can be the case that factoring 1 side is what you want. This won't, of course, change that side's value, but it will **re-express** the equation in a way that might be useful. This is obviously the case with quadratics, which are often solved through factoring, but it can be helpful in other places too. If you see $4x - 12$, it might be advantageous to rewrite it as $4(x - 3)$. That might look like you're **complicating** things, but it could be just what you need to get to the next step.

Tip 7

SPR. These are student-produced questions, the ones where you enter the numbers yourself. Make sure you know how to enter fractions and decimals. You're limited to 5 characters, including decimal points and fraction slashes, unless it's a negative number, in which case you can enter 6 characters. If your answer is $1\frac{2}{3}$, you **cannot** enter it as 1 2/3. It's 5/3 or 1.666 or 1.667.

Nerves

It would be a little odd if you didn't feel a bit nervous about taking the actual SAT. The simplest fix is to know the material as well as possible. Even if you consider yourself a not-so-good test-taker, you wouldn't be stressed taking a test where you had to add single-digit numbers, right? You've mastered that.

And by now, you know much more about the exam than you did a few weeks ago. You know more about it than most people. You might not feel you've mastered the material, but taking these practice tests will solidify your knowledge and allow you to find those areas that still need work. If you make a mistake, tell yourself, "Yes! Here's a question I would have gotten wrong on the real test but can now fix." It's a golden opportunity.

Of course, on the day of the actual exam, it's possible that you'll get a little stuck now and then, unsure of what to do next. Sometimes the best thing to do is to **stop taking the test**. For a few seconds, look away, take a breath, maybe stretch your arms and neck or whatever, and then try again. These micro-breaks can really help. If you're still stuck, mark the question so you can go back to it, eliminate any choices you can, and move on.

It's also worth keeping in mind that the test can be taken multiple times, and the best score you get on each section can be combined into a **superscore**. So unless you're a senior taking the test in the late fall or early winter, there's a good chance you can take the test again and push that score up some more.

Day 24

Practice Test 2

After reading this material and completing Test 2, sign and check the box for Day 24.

Today is when you get to try out your new skills under timed conditions with Test 2 from Bluebook, the official online test application. Tomorrow you'll review the questions. This "take-a-test/review-a-test process" will continue for Tests 3 and 4. We strongly advise that you also complete the two Reading and Writing modules on *at least* 1 of these days. You'll want to experience what it's like to do the Math modules *after* you've sat for the other ones. And of course, even if you're not too concerned about the Reading and Writing section, it's still a good idea to practice this material and investigate any mistakes you've made. But you might remember this from Day 1:

> *If you are only taking the Math sections, you'll need to bypass the two Reading and Writing modules to get there. The easiest way to do that is to skip every question on those modules so it will take you to the Math section. Yes, it will look like you didn't get a single question right on the Reading and Writing section, but no one will ever know.*

So find a place such as a library where you won't be interrupted. Make sure you have a few pieces of scratch paper and a pencil handy. Power off your phone. Expect to occasionally get a little stuck, but don't panic. If you need to take an intelligent guess after eliminating a choice or 2, that's okay.

By the way, each module on the actual exam will contain perhaps 2 questions that will not count toward your score. That is also true for the official practice tests. These questions are only for research purposes to test out the questions for the test authors. Of course, you'll have no way of knowing which ones they are, but if you do get in a little trouble, you can always tell yourself, "Maybe this is one of the questions that don't count!"

And then move on.

Day 25

Reviewing Practice Test 2

After reading this material and completing your review of Test 2, sign and check the box for Day 25.

There should have been no surprises on this test, although you might also have gotten a bit of a reality check working under strict time guidelines. When reviewing the material, ask yourself, "How could I have gotten that question right?" and also "How could I have gotten that question right (which I did) more efficiently?"

If you got a question wrong due to carelessness, don't just kick yourself and move on. Think carefully about how you can avoid that carelessness in the future. It doesn't matter **how** you got it wrong. A wrong answer is wrong for whatever reason.

There are plenty of quadratics on this test, of course. A question involving the volume of a cylinder, which hopefully you know without consulting the reference guide, appears in the first module. Some of the later and (perhaps) tougher questions can be handled smoothly with the OSC. Did you use it to your advantage?

A little trig appears, as always, and there's no reference guide for SOH CAH TOA, so you need to know that material **cold**!

The test folks provide explanations for every question, online and in their book, but again, we recommend the explanations you'll find on Khan Academy. You can find a link to it once you have completed the test.

When you've mastered this test, you're on to Test 3.

Day 26

Practice Test 3

After reading this material and completing Test 3, you can sign the check the box for Day 26.

By now, you should have done a thorough analysis of Test 2. Perhaps you went back to some of the earlier material in this book to review some topics. Did you find areas that need sharpening? Did you locate some questions you could have gotten right in half the time? Are you getting a better sense of when to and when not to use the OSC?

We'll just assume you said yes to those questions, so great! Now it's time to apply all that to this next practice test. If you haven't yet taken the Reading and Writing portion—or even if you have—consider doing so for this Test 3.

Day 27

Reviewing Practice Test 3

After reading this material and completing your review of Test 3, sign and check the box for Day 27.

On this test, you'll see a lot of lines. That's nice because you know all about them—slope, intercepts, simultaneous linear equations—and there's not much more to them than that. How did you do with the box plot question? It's not at all difficult if you're familiar with them. Did you know how to go from degrees to radians, and vice versa? Obviously, you should! (Remember, 180 degrees is π radians.) Did you find yourself using the OSC now and then? You can't tackle every question using it, but when you can (and you **should**), **DO!**

Again, the test people provide explanations, but they're frequently overly long. You'll almost certainly do better consulting Khan Academy to review specific questions and then this book for concepts and drills.

There's 1 more practice test to go. Make sure you completely "get" all the questions on Test 3. You're sure to see the same material on Test 4 and, of course, on the real thing!

Day 28

Practice Test 4

After reading this material and completing Test 4, sign and check the box for Day 28.

You've now been exposed to anything that might be on the actual exam. Your feel for the timing should be solid. Make sure that today you're in a place—and a headspace—that prevents any interruptions. Mimic the test conditions as best you can. If you're also taking the Reading and Writing section, give yourself the full break before the Math modules. Use that time to stretch and, if you want, have a snack and some water.

You'll find out today and tomorrow whether there are still some areas that need polishing. If so, do what you need to polish those areas to a high gloss.

You're almost there!

Day 29

Reviewing Practice Test 4

After reading this material and completing your review of Test 4, sign and check the box for Day 29.

Averages, similar triangles, simultaneous equations, the perimeter of squares, absolute value, inequalities, roots with exponents, translated lines, the circle equation—these are nothing you haven't seen before. How was your timing? Did you take any intelligent guesses and then move on a few times? Sometimes that's the right thing to do.

Remember that the goal is not to get every question right—unless you believe a perfect score is within reach—but to get the most questions right within the time allowed. Sometimes that means giving up gracefully, **consulting the choices**, knocking off 1 or 2 of them if you can, taking a guess (no, it doesn't matter what letter you choose at that point), and moving on.

Tomorrow we'll review some of the most important content, strategies, and time-management tips. For most people, this last day is not a time to tackle a bunch of new material. We'll take a step back, look at what's most important, perhaps close a gap or 2, and make sure we're ready in every way for the big day!

Day 30

Course Review – Big Picture, Details

After reading this material and completing the drill at the end, sign and check the box for Day 30.

Whew!

That was a lot of material to get through in a mere month. You should now feel equipped to handle anything the actual exam sends your way. You might even have helped others learn the material, which is actually a **great way** to solidify your own knowledge. If you can explain a difficult question in a concise, comprehensible manner, you really understand it. It's best not to tackle anything too new today, but why not look at some of those last few math questions in each module and see if you can— out loud—explain how to solve them. You don't need an actual human to listen to you. Your cat, dog, salamander, or a stretch of wallpaper will do fine.

The Big Picture

The point is— as it always has been—**efficiency**. How smoothly can you move through the steps required by a problem? Can you reduce it from 7 steps to 4? If you're solving an algebraic equation, can you write out the steps in an orderly manner that reflects your orderly thinking? Your pencil is your brain's assistant. If it's jotting down numbers and letters in a haphazard manner, it's not doing a good job of assisting. Neatness isn't exactly what we're getting at, but **organization** is the goal.

Take a question from the course or from the test that took you a while to plow through (whether you got it right or wrong), and see if you and your pencil can make your way from start to finish in the most efficient path possible. Then, perhaps try it again. Not many questions on the test really require a long drawn-out process.

But also remember another assistant at your command—the **OSC.** Are you looking for the intersection of 2 lines, a line and a parabola, or maybe a line and a circle? That's where the OSC shines, especially when you have just 2 unknowns.

And you have **4 other assistants**—the answer choices. If you're stuck, and you've heard us say this before, **consult the choices**. Too many students give up without doing that. Perhaps they glance at the choices but don't truly consider how they might help them. Can you knock off a choice or 2, or even 3? Are any of the choices too obvious? Can you backsolve from the choices?

Of course, you can't rely on choices for help on the SPR questions. For these questions, which appear at random points within each module, you supply the answer yourself. They're not tougher than the other questions; the format is just different. Make sure you understand how to enter the numbers (see Day 23). And just as with multiple choice questions, **never leave it blank**. If you have to guess, go for it.

Details

Reviewing **Quizzes 1, 2, and 3** is a good way to see how you've progressed and what else you can still understand a little—or a lot—better. Besides that, let's do an overview of the key elements of the course, starting with the first real lesson on Day 2.

Day 2: We started with a review of basic arithmetic, with an emphasis on fractions, exponents, and roots. That's because much of the exam requires you to manipulate these elements in various ways. If you have any uncertainty about cross-multiplication or negative or fractional exponents, have another look at this lesson. You **cannot** let these basic steps slow you down.

Day 3: Handling simultaneous (multiple) equations is something you're absolutely going to need to do on test day. How are you at substitution, elimination, using the OSC, and choosing among these methods for the most efficient approach? And don't forget this key rule regarding **in**equalities: flip the sign when multiplying or dividing by a negative value.

Day 4: The test people seem to think you can't be a success at college unless you can handle **quadratics**. These are all over the test. Can you solve through factoring? Can you group? Can you complete the square so you have the vertex form of a quadratic? Do you know the quadratic formula? Can you enter a quadratic into the OSC efficiently? This lesson is a biggie.

Day 5: You need to know the rate and average formulas in their different forms such as "Sum = Average x Number of terms." It's important that you know when to consider **backsolving** (when the choices are known values) or **picking numbers** (when the choices contain unknown values such as x).

Day 6: Know your area and volume formulas. Know *especially* the importance of right triangles, *and* know the special triangles such as the 45/45 and the 30/60. Other people will know these; you need to know them **better**.

Days 7 and 8: Quiz 1 and Explanations

Day 9: Mastering this lesson on coordinate geometry will pay off considerably on test day. There's a relationship between algebra and geometry that is spelled out here.

Day 10: The test folks want to know whether you can solve an equation but also whether you understand what the numbers—known and unknown—represent in an equation or expression. And if you don't quite get compound interest, which can show up in various ways on the test, have another look at this lesson.

Day 11: There's not a lot of trig on the exam, but knowledge of the basics will be a huge help on some of the tougher questions. Are you still shaky about radians? This lesson will help.

Day 12: You need to be comfortable with the various ways that statistics like mean and standard deviation are represented graphically. And you should be comfortable with calculating the probability of multiple events. If you're not, have another look here.

Days 13 and 14: Quiz 2 and Explanations

Day 15: Break

Day 16: This lesson is focused on the trickier algebra that usually appears toward the end of a module. How efficiently can you handle a multi-step algebraic equation? Are you comfortable with the concept of an equation with no solutions, as well as with infinite solutions?

Day 17: This lesson looks at how trig and geometry can be tested in the same question. It also looks at arcs and translations. You might get 1 arc question, and perhaps 1 translation question. Isn't it great to be prepared for it?

Day 18: You'll almost certainly see some "miscellaneous" material on the exam such as sequences and statistical inferences, though probably not all of it. Be ready for anything!

Days 19 and 20: Quiz 3 and Explanations

Day 21: You looked at this lesson about the test experience not long ago, but it's not a bad idea to have another look at it. Do you have everything you need for test day?

Day 22: Nothing here is directly related to the test since it's about admissions.

Day 23: Sure, you looked at this lesson recently, but it's good to have another look at it since you're so close to the actual exam. Do you know all the time-saving tips?

Days 24–29: These are practice test days. Do you think there's anything here worth looking at again, even if you've already looked at it 2 times, 3 times, or more?

Day 30: And that brings us to today.
We don't want to say much more here because it's better that you spend your time in the way you think best. But here are a few things to keep in mind going into the exam.
1. It's not a race. The student next to you might finish before you, and that might be because they were moving so fast that they quickly determined that $7 - (-1) = 6$. (It's not.)
2. Don't lose sight of what the question is actually asking. This seems obvious, but when you're stuck, re-read the question.
3. Many of you will take the test again and perhaps more than that. This is a good idea, especially since so many schools "superscore" your various test scores. This book can be immensely useful to you.

4. If you're going to take the test again, for a little extra money you might be able to order a copy of your exam with your answers. However, this service is not always available. If it is, we highly recommend that you do it.

5. Perhaps you're considering taking the ACT at some time. Everything in this course applies to that exam, although currently the ACT is not adaptive and does not have an OSC. Even though all the test content on these pages will help you with the ACT, some topics on that exam (imaginary numbers, for example) are not found on the SAT and therefore not in this book.

Finally, know when and where you'll be taking the test. Get to know the parking situation. And know the equation that defines a circle.

You're ready.

Now that you have everything you need to master SAT Math, it's time to pass on your newfound knowledge and show other students where they can find the same help.

Simply by leaving your honest opinion of this book on Amazon, you'll show other students where they can find the information they're looking for, helping them with their own SAT journeys.

Thank you for your help and for supporting this learning community!